MYSTERY ANIMALS OF THE BRITISH ISLES:
The Scottish Borders

Tom Bryan

Typeset by Jonathan Downes,
Proofing by Ve Macrinnon, Karen Heath
Cover and Layout by SPiderKaT for CFZ Communications
Using Microsoft Word 2000, Microsoft Publisher 2000, Adobe Photoshop CS.

First published in Great Britain by CFZ Press

CFZ Press
Myrtle Cottage
Woolsery
Bideford
North Devon
EX39 5QR

© CFZ MMXXIII

All rights reserved. Without limiting the rights under copyright reserved above, no part of this publication may be reproduced, stored in or introduced into a retrieval system, or transmitted, in any form of by any means (electronic, mechanical, photocopying, recording or otherwise), without the prior written permission of both the copyright owners and the publishers of this book.

978-1-909488-68-7

Dedication: To my beloved wife Janet Elisabeth "Lis" Lee (1948-2021) With special thanks to my brother Garry who has devoted his life to the mystery and natural wonder of all animals, especially less popular and venomous ones!

This area comprises the current Scottish Borders Council governance.

Contents

GEOLOGIC TIME: FOSSILS AND TETRAPODS
INTERLUDE: EXTINCT ANIMALS
SPIDERS
INSECTS
INTERLUDE: FRESHWATER PEARL MUSSEL
FISH
AMPHIBIANS
REPTILES
INTERLUDE: GIANTISM
INTERLUDE: THE GIANT SQUID
BIRDS
RELICT ANIMALS
OCKHAM'S/OCCAM'S RAZOR
MAMMALS
INTERLUDE: ANIMALS IN THE BORDER BALLADS
DANDIE DINMONT
ZOOFORMS
BIBLIOGRAPHY

APPENDICES
The Mystery Animals of the British Isles (Jonathan Downes) Jonathan's own introduction.

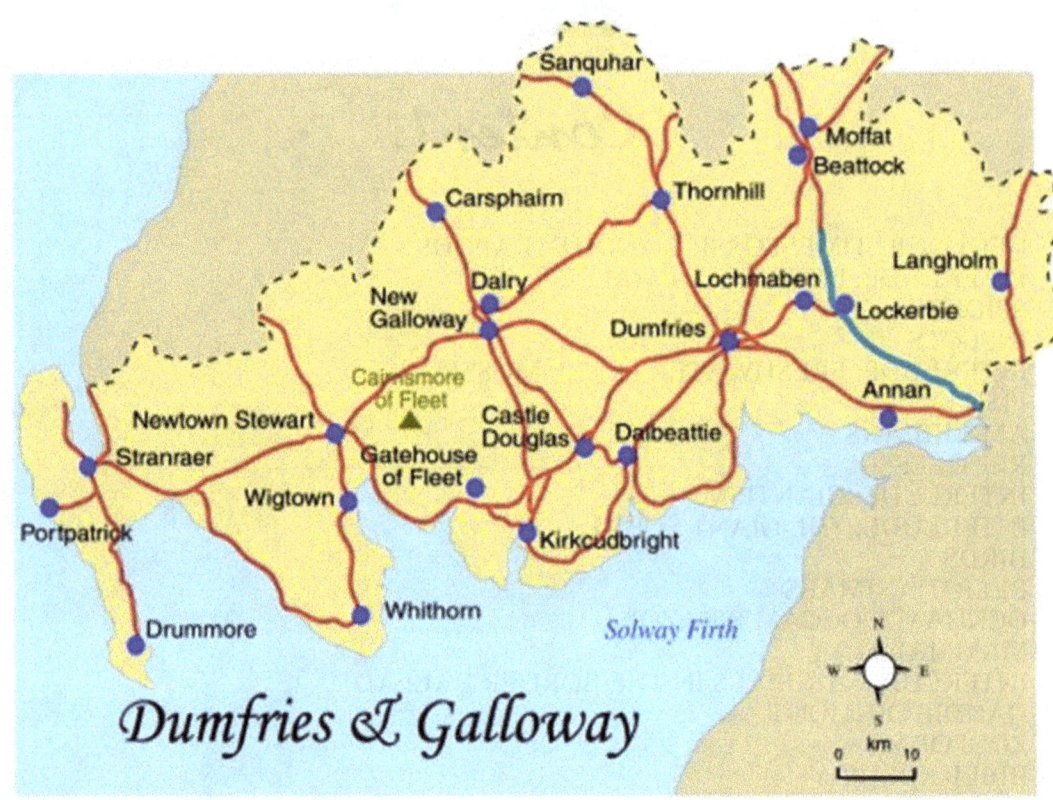

Dumfries and Galloway comprises the counties of Dumfriesshire, Kirkcudbrightshire and Wigtownshire (from east to west). The southern shore of the Solway Firth is Cumbria/Cumberland.

Introduction

I've lived nearly a third of my adult life in the Scottish Borders. Some of my ancestors lived here for hundreds of years before emigrating to Canada and America. Traditionally, 'Borders' has meant the border with the English counties of Cumbria/Cumberland and Northumbria/Northumberland, where my English forebears also go back for centuries. It comprises the current area of the Scottish Borders Council: Berwickshire, Peeblesshire, Roxburghshire and Selkirkshire. The traditional Borders also covered areas of South Lanarkshire, Midlothian and Ayrshire but chiefly, this book includes the area governed now by Dumfries and Galloway, comprising Dumfries (shire) Kirkcudbrightshire, Wigtownshire and Galloway.

In the east there is a small sea coast with Eyemouth as its gateway. In the west, the Solway Firth and Irish Sea offer a much more extensive seascape.

This book is thus chiefly limited to those counties bordering England, with only the occasional 'liberty' taken with the fluctuating boundaries!

The Borders has a rich history: Thomas the Rhymer, the Ballads, Sir Walter Scott, and James Hogg. Many songs and legends tell of the Reivers (Mosstroopers), those bold men who rode out from both sides of the Border, existing by cattle raiding and martial codes of both honour and expedience.

I am neither a zoologist or animal specialist, though have written and researched as an amateur with a keen interest in all wildlife, particularly fishes and insects. I am also a published poet and writer who finds in the Borders a perfect blend of fact and fiction, and of truth and imagination. This book deals mostly with vertebrate animals because that's what I know best: both real and imaginary. When Jonathan Downes suggested a book for his series I thought it was a book I wanted to do and could do.

The book's purpose is to remind us what we have and resurrect knowledge of what has been lost. As a species we need to both protect and nurture the wildlife we have yet honour what we have lost through stupidity or carelessness.

I hope this book gets people outdoors to wonder at the mystery of all living creatures.

Mystery Animals of the Scottish Borders

The Borders is a great natural landscape: rivers, lochs, hills, forests and seashore. It has remarkable diversity in a (relatively) small area.

Like most people living here, I think the Scottish Borders does not get the recognition it deserves, perhaps being eclipsed by Edinburgh to the north, the Lake District to the south, as well as Scotland's own stunning Highland landscapes.

I would like you to have some fun with this book. It is a work of the imagination and as human beings with a short time on the planet, we can put our imagination to good global use. Knowing why something has disappeared might prevent us from repeating our past mistakes. Very real animals (including very small ones) need our support whilst more imaginary ones perhaps at least deserve our wonder. 'Mystery Animals' is almost a tautology: life is full of mystery, from the smallest insects to the largest predators. Mystery here means animals that *are, that were and that can be imagined.*

For years, I was a full-time carer for my late wife and thus was not able to stravaig the Borders as we both once did. I ask for some forbearance on that score and apologise for any omissions; this book is a taster and introduction rather than an encyclopaedia or compendium. Many dedicated (mostly unpaid) Borderers are already working very hard cataloguing, describing and preserving the wildlife that does exist, maybe without the luxury of my own imaginary flights. Ultimately, this book is only one man's view. I can accept that this book is both eccentric and unusual and that can be condemnation in some circles, but praise in others.

When out in the countryside, please respect the Countryside Code and do not disturb or intrude on wildlife. A camera is essential and a notebook (waterproof ones are available) is also a must. Try not disturb the natural habitat and always be respectful. Be able to report clearly what you see to trained naturalists or animal experts, without prejudice. Keep an open mind on what you are seeing or think you are seeing. DNA evidence may be vital (spoor/fur/hair/scales/skin); this can often be obtained without harming wildlife but please use your common sense here.

In March 2020, the Corona Virus/Covid-19 first appeared in the Borders and the restrictions placed on travel and mobility resulted in a landscape that was suddenly quiet and relatively free of human activity. However, it also encouraged some people to go outdoors and perhaps pay more attention to nature. Many people heard owls for the first time that had previously gone unheard because of traffic noise. Garden birds were seen in greater numbers and of course, less traffic meant less roadkill. As of this writing (February 2022) variants of the Covid virus are still causing illness and some death in the Borders but it appears the epidemic is lessening. The long-term effects on our wider environment are yet to be determined but we have learned that people are safer outdoors than in so that may also encourage us to take more delight in our natural world.

Tom Bryan

Foreword

Something in excess of thirty-five years ago, back in a completely different life, my first wife - Alison - bought me a book for a birthday present. It purported to be about the "mystery animals of the British isles" and although it opened a whole bunch of new doors for me, most specifically introducing me to the life and work of the legendary Tony 'Doc' Shiels. It was in other ways quite a serious disappointment, because even then, as a cryptozoological novice who had been interested in this stuff ever since the age of eight, I knew that the Natural History of the British Isles was full of amazing and often unresearched lacunae. I had hoped that this book would be the same, but it wasn't.

Sure, there was a chapter full of bits and bobs from various places in the canon of Fortean literature, such as 'The Borley Bug' and a bunch of less tangible creatures, but there was a whole lot that I knew had been left out.

This is why I started this series for CFZ Press something like fifteen years ago. I wanted a bunch of books, each dealing with a specific geographic area of our islands, and containing the sort of information that nobody living outside of that specific area would ever have heard of. I hand-picked authors that I knew for as many of the regions that seemed appropriate, but that only took up something like fifteen or twenty percent of the British Isles. I was approached on occasion by an author of whom I had never heard before, and I had to decide whether to take a punt on them, purely on instinct and sometimes a few sample chapters.

And so it was with Tom Bryan.

Not only does he write in a telling and engaging fashion, but although I had been to various parts of the Scottish borders in my various capacities over the years, it was only when I first read the original manuscript of this book that I realized how little I really knew about the area: It is with hand on heart that I can state that this is the sort of book that I had so badly wanted to read on my birthday all those years ago.

I particularly like the way that he intersperses modern day sightings of creatures which have allegedly been extirpated from the region with the sort of in-depth look at creatures like the Arctic charr (the cryptid status of which I will freely admit interests me far more than yet another bunch of "big cat" sightings or claims of "bigfoot", where such a thing could not possibly exist). It is his in-depth look at anomalous creatures like these which most authors on the broad subject of mystery animals would not even think about covering that really makes this book for me.

I would like to congratulate Tom for having authored such a fantastic little book and despite unavoidable delays caused by the tragic deaths of both our wives, and I am very glad that we finally got it together enough to publish this long-awaited book.

It has been a long time coming, but I am very excited with this book and I hope that you will be too.

Jonathan Downes
(Director, Centre for Fortean Zoology)
Woolsery
North Devon.

February 2022

Mystery Animals of the Scottish Borders

Mystery Animals. This book is part of a series of "The Mystery Animals of the British Isles". This series gives wide scope to writers to fill in that mystery. Some have chosen a more fanciful approach whilst I have tried to strike a balance between fact and fiction. Here are a few useful definitions:

Cryptid: Unknown species of animals or known species of animals awaiting rediscovery or verification. These animals may have been incorrectly thought to be extinct.

Cryptozoology: Bernard Heuvelmans definition: "The scientific study of hidden animals, i.e., of still unknown animal forms about which only testimonial and circumstantial evidence is available..."

Melanic/Melanistic: Although melanic is the accurate dictionary term, melanistic is used more often. This describes a dark or black version of a species that is not normally dark or black. This condition is often rare but more common in some species than others. For example, melanistic cougars are not known to exist but black jaguars and leopards have been noted. Black panthers, seen most often in zoos or films, are usually melanistic versions of leopards, less often, of jaguars.

Geologic Time: Fossils and Tetrapods

Sites in the Scottish Borders are some of the most important in science at the moment because of mysterious creatures that are no longer with us but have left traces of their life. Tetrapods are four-limbed/legged vertebrates. This big family of animals includes frogs, turtles, newts and salamanders as well as mammals like rats, rabbits and squirrels. Probably the most interesting tetrapods are those that changed from aquatic animals (like lobe-finned fishes) to a species that could also survive on land like frogs, crocodiles and turtles. Moreover, these scientific discoveries in the Borders are helping solve one of palaeontology's (the study of fossils) greatest mysteries: why are there such long periods without a fossil record, particularly a record of early tetrapods: our most important land-dwelling ancestors.

A lot of the mystery is down to the work of one man: Alfred Sherwood Romer (1894-1973) Romer (right) was a distinguished palaeontology, professor and Museum Director. Romer puzzled over the

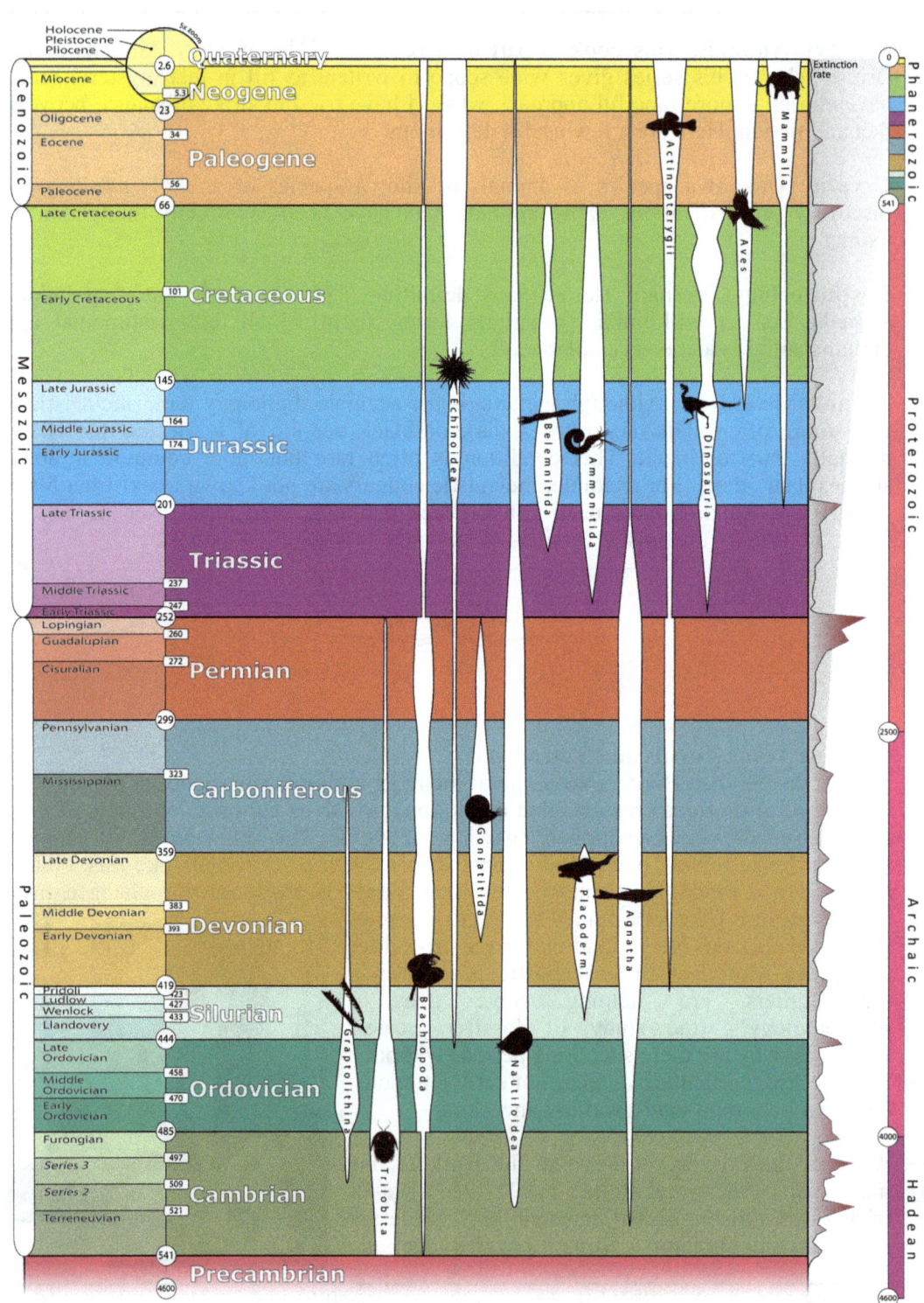

absence of key fossils from the fossil records which would give us a much greater knowledge of that important evolutionary step from water to land. Since 1995, that important fossil gap has been known as 'Romer's Gap'. Many outstanding researchers and field workers have tried to bridge that gap in the Devonian and Carboniferous periods. This fifteen million year gap is finally being bridged, in some key sites in the Scottish Borders, in as many as five fairly secret locations. In 2012, 350 million year old tetrapod fossils were found near the Whiteadder Water near Chirnside as well as the Tweed near Coldstream on the English border. One of the key finds was an early amphibian nicknamed 'Ribbo'

A press release from the University of Leicester Press Office on 16 February 2016 explained the importance of the discoveries in the Scottish Borders. I quote:

"Until recently, there was very little fossil evidence of life on land during the early Carboniferous period, around 345-360 million years ago. This was a pivotal moment in evolution, when vertebrate life moved from the sea to the land; a momentous shift without which humans would not exist today.

However, the fossils found in the Scottish Borders confirm that a rich and diverse ecosystem of amphibians, plants, fish and invertebrates thrived during this period...

The late Stan Wood, a self-taught Scottish field palaeontologist, was convinced that fossil evidence of tetrapod life during 'Romer's Gap' could be found in Scotland and spent twenty years searching for it. Finally, in 2011, he uncovered never-before-seen fossil evidence of early tetrapod life on land-fossil animal skeletons, along with millipedes, scorpions and plants-at the Whiteadder River near Chirnside."

This is revolutionary work but also illustrates the value of an alert self-taught field worker. All of us can lend our eyes, ears and observation to science. We must always use our imaginations as well as our usual senses. Imagine this: our Scottish Borders was at the equator when these creatures lived. It was covered with huge rain forests in a hot and humid climate, perhaps more like a jungle. It was one of Scotland's first wetlands. Of the five key tetrapods found (bones, fragments and skull fragments) my favourite is *Aytonerpeton microps* ("small-faced crawler from Ayton") a small village near Burnmouth.

These exciting discoveries will continue and perhaps 'Romer's Gap' may finally be bridged.

The only caveat here is to leave the sites and work to the professionals, even self-taught ones. We can't see 'RibbO' or the "small-faced crawler from Ayton" but we can imagine them. Mystery animals include real ones as well as mythical ones, even if the real ones lived 365 million years ago!

Interlude:
EXTINCT ANIMALS

Part of thinking about mystery animals is surely also thinking about why so many wonderful species have disappeared from our islands, some in fairly recent times. Of course, extinction is a global phenomenon. This disappearance is to our guilt and shame and should make all of us determined to protect the wildlife we have, from the smallest insects and arachnids up to tigers, whales and rhinos. Worldwide, some claims for unidentified creatures are based on the assumption that an unknown animal in historical times might actually have survived undetected to modern times. For example, very few writers agree when wolves became extinct in Scotland and Ireland. We don't have coelecanths in Scottish waters but its story is instructive for cynics who may think *'extinct means extinct'*.

The coelacanth (*Latimeria chalumnae*) was a lobe-finned fish which first appeared over 350 million years ago. Its range was extensive and fossils for the species had been found all over the world. It was assumed to have become extinct with the dinosaurs about 65 million years ago. However, in December 1938, a five foot long, 127 pound blue scaled fish turned up in a South African fish market. A museum curator and taxidermist named Marjorie Courtenay-Latimer was in the market that day looking for fish for her museum collection. She later identified the fish as a real coelacanth, a 'living fossil'. Unfortunately a taxidermist had destroyed the organs and tissue that would have confirmed her identification. However, finally, another confirmed specimen turned up off the coast of Mozambique in 1952. Since then, more than two hundred coelacanth specimens have been caught and identified. The real lesson is that local fishermen had been catching and eating the fish for decades previously. Science had thus lagged behind local knowledge and it is not impossible that we will yet discover 'extinct' animals in Scotland or Britain generally.

'Nessie' is assumed by some to be a plesiosaur or zeuglodon. Woolly mammoths are thought to still exist in the Siberian forests. It might be useful to look at some of the vertebrate animals that have vanished from our Isles and may yet account for some of the unidentified animals seen in the Borders, in Scotland and Britain as a whole. For example, many of the mysterious Big Cat sightings in the Borders in recent times are described as 'lynx-like' yet the Eurasian Lynx is thought to have become extinct in Scotland as early as 1250.

Most of our extinct animals are exquisite beetles, moths and arachnids, not included here only because their identification is highly specialised. However, any sightings of unusual specimens should be photographed, with as many details noted as possible. It is more likely that these micro-creatures have escaped extinction only because they have escaped detection. There are many success stories of smaller creatures being rediscovered, nurtured or reintroduced with improvement in their natural habitat. In order to know what we might see, it is good just to review some of the large vertebrates that may have been familiar to our ancestors in Scotland.

European Brown Bear: 10th century
There are various estimates about when the European Brown Bear (*Ursus arctos*) disappeared from Scotland. The Roman poet Martial makes mention of a Roman gladiator who was mauled by 'a Caledonian bear'. Bear teeth have been found in Caithness and a well-preserved skull and rib were found in a Dumfriesshire peat bog at Shaws. Some argue that Clan Forbes takes its name from the supposed slaughter of a bear by a chieftain who then became 'Forbear' or 'Forbeiste.' The traveller Pennant refers to some of Clan Gordon slaying a bear in 1057. The great folklorist John Francis Campbell recorded tales of bears in the West Highlands (*Popular Tales of the West Highlands*) and I recall reading that Campbell once met an elderly Gaelic crofter who claimed to have encountered bears in his youth in Sutherland! Campbell lived from 1821-1885 and compiled his *Popular Tales of the West Highlands* in 1860-1862, thus Campbell's Gaelic-speaking informant would have been a young man in the late 18th century. He supposedly responded to an illustration of a brown bear in a book of Campbell's. However, all evidence indicated that the brown bear may have disappeared from the Scottish Borders in early medieval times and was certainly well gone by the 11th century.

European Beaver (*Castor fiber*) 1200
Beavers were present in mainland Britain from the Ice Age, having crossed land bridges from elsewhere in Europe. They were hunted to extinction for the value of their pelts, far more desirable to hunters than furs of martens or otters. Beavers became extinct in England in Saxon times, in Wales by the end of the 12th century and in Scotland in the 16th century. Although beavers have been reintroduced in Scotland, there are also thriving colonies of escapees in the River Tay. As these populations expand and increase, any reported sightings of beavers in the Borders river systems might be credible. These are intelligent and adaptable animals that will seek new territories.

Here is an incomplete list of other vertebrates than have become extinct in Scotland, with an approximate date of when that happened.

Mammals:
Eurasian Lynx 1250
Wild Boar 1263
Grey Wolf 1786
Greater Mouse-Eared Bat 1988

Birds:
Capercaillie 1769
Common Crane 1800

Great Bustard 1832
Great Auk 1844
Black Tern 1884
Northern Goshawk 1893
White-Tailed Eagle/Sea Eagle 1918

Fish:
Burbot 1969
Amphibians:
Pool Frog 1995

Mystery Animals of the Scottish Borders

SPIDERS

Scotland has at least 400 species of spider. Many can only be identified by experts with microscopes. The Wildlife Information Centre has an excellent website for anyone wishing to do spider research:

www.wildlifeinformation.co.uk/scottish_spider_search.php

That organisation recently asked for help with more field information about four spiders found in Scotland, if not in the Borders.

1. Four-spotted Orbweb Spider
2. Daddy Long-legs Spider
3. Zebra Spider
4. Nurseryweb Spider

Likewise, please consult the website for the British Arachnological Society: **britishspiders.org.uk**

The most valuable information about spiders can be gathered by field workers. Please consult both websites about the best way to help them gather that information.

The Giant House Spider

Just a regression: although I've seen tarantulas in zoos and as private pets, the biggest spiders I've ever seen were on a friend's ceiling near Melbourne, Australia. There were three of them (motionless) as big as my hand. "Don't worry" he said, "the Huntsman Spider is harmless and generally avoids people." Not comfortable sleeping in a room with them just above my head, we tried in vain to catch them but my heart wasn't in it. There is a sequel to this story. A friend in Kelso had visitors from Australia who were familiar with Huntsman and other big spiders. However, they were pretty impressed (and a bit frightened) of a local spider which they thought at first was a mouse running along the skirting boards! It was a Giant House Spider which we explained didn't often use its web because it could out-run its prey. Many residents on Roxburgh Street reported similar: these house spiders were big and fast, bigger and faster than described in the textbooks.

A sequel to this occurred later in the week (September 2004) Near Kelso Square, a few witnesses (including my fiancée, later wife) saw a gigantic spider walk across the street in the direction of the Royal Bank of Scotland. It vanished somewhere on that side of the street. My guess is that it might have been an escaped pet tarantula. She joked that because it was heading for a bank it was a true 'money spider'!

Some years after this, local employees of a firm near Glasgow had safely captured a Huntsman Spider on their premises, after a shipment of goods from Australia had been received. Another possibility is that the huge Kelso spider had come in as a stowaway in one of the local supermarkets, not unheard of elsewhere in Britain.

I'm not making a claim that Kelso spiders are bigger than those elsewhere, just that it takes some doing to impress Australians with Scottish spiders. I expect that with the proximity of field, woodland and river close by, there is an ample food source to help the spiders grow bigger and faster than the average. The spider seen on a busy street in Kelso is yet another example of expecting the unexpected, a good rule for any observant naturalist.

Mystery Animals of the Scottish Borders

INSECTS
A Strange find in Peebles

Peebles is in the more rugged west part of the Scottish Borders Council counties. It is a lovely village on the River Tweed and the landscape west of Peebles changes dramatically, resembling parts of the Highlands. If you approach Peebles by road from the east, as I have done hundreds of times you come to a roundabout. If you take the turnoff to Edinburgh, after less than a mile, you will pass Dalatho Crescent on your left. My great uncle lived there in the Fifties, from where he took the bus faithfully every day to work as a dental technician in Edinburgh. One day was a bit different. He got off at his bus stop and arrived home to find his wife gesturing excitedly to him. "Look at this" she said, taking him to the gable end of their house. They both wondered at what they saw. It was a brightly-coloured moth, larger than my great uncle's working man's hand. They put the creature in a shoe box and probably spent much of the evening checking on its condition. Unfortunately, the colourful insect died. However, my uncle's practical mind took over. He took the moth to work the next morning and at lunch time, went to the museum at Chambers Street. He asked for the man who "does moths" and simply told the man his story and handed him the box. The man opened the box indifferently but his indifference left him. "You may have found this in Peebles" he said "but it is one of the larger more colourful moths in the world, found primarily in the Atlas Mountains of Morocco ."

For many years, the moth had a pride of place with a brief note of its strange history. My relative couldn't remember its name but could easily recall its many bright colours and its unusual size.

INSECTS: IN THE FIELD

What is it? **Conformist Moth** (*Lithophane furcifera*)

Most of the exciting work done with 'mystery animals' is done by dedicated people who must study what most of us can't even see: the world of micro-creatures. The Conformist Moth is not a micro-creature but it is elusive and requires a great deal of research.

What is its natural range?
It was last recorded in Glamorgan, Wales in 1959 and recorded in various parts of England until 1946. It was recorded in neighbouring Northumberland in 2007 and in Scotland in 2011, at the RSPB Insh Marshes Reserve in Strathspey.

Mystery Animals of the Scottish Borders

Where has it been seen in the Scottish Borders?

I can't prove that it has ever been seen in the Scottish Borders but given the nature of moths and their reproduction, if it has been recorded in Northumberland then it can't be ruled out in the Borders either.

Where to look for it?
The larvae feed on elm, oak, willow and poplar.

What to do?
First, before doing any fieldwork, do some research on this creature. Find out which individuals are experts in this field. Do the same with any clubs or associations in your area. Talk to them first, before taking to the field. Find out from them what you can do to help.

Small Blue Butterfly: *Cupido minimus*

This species is one of the smallest in Britain. It is usually associated with the chalk and limestone country in the south of England but isolated Scottish populations have also been found along North Sea coastal areas, possibly including coastal areas of the Solway and of the North Sea. These small butterflies only live for about a fortnight and often lose their colour and camouflage during that time as their wings lose scales and become ragged and useless. Their tiny caterpillars only eat the kidney vetch plant so these plants should be the basis of any search. The Small Blue is found in Cumbria so naturally specimens could cross over into Scotland. Fortunately, there are many knowledgeable naturalists in the Scottish Borders keeping an eye out for this lovely butterfly. It is a protected species so a good photograph could be invaluable as well as specific sighting details and map location. Under no circumstances should the insect be disturbed or handled.

Because kidney vetch (*Anthyllis vulneraria*) is the sole food plant of the Small Blue's caterpillar, it is useful for the searcher to know about this plant. Although it can grow all over Britain, it prefers dry grasslands close to salt water. Because its flower-heads are kidney-shaped, it was thought that the plant could be used to treat kidney diseases and heal wounds. More confusing to those in the field, the kidney vetch has a variety of local dialect names throughout its British range: *butterfingers, fingers and thumbs, double pincushion, lamb's foot, lady's finger and many more.*

It also boasts a bewildering range of flower colours: yellow, orange, red, purple and shades of off-white. Its leaves and stems are silky.

The plant also has a rich supply of nectar, which is so difficult to extract that only the strongest bumblebees are able to manage it.

Insect
What is it? Pincer-tailed Caddis Fly (*Hydroptila igurina*)

What is its natural range?
It is a river-dwelling insect. It is found in Europe, in stony, fast-flowing rivers.

Where has it been seen in the Scottish Borders?
It has never been recorded in the Scottish Borders but was last recorded in Ambleside, Cumbria in 1881. However, in March 2010 it was discovered in Assynt, Sutherland, Scotland. This has been its only Scottish recording. However, experts believe that it can also be found in the wide range between Cumbria and Sutherland. Much of the Borders would present possibilities because of the prevalence of ideal habitat for the Caddis Fly in the Borders.

Where to look for it?
Stony, fast flowing rivers: the Tweed, Ettrick, Yarrow, Teviot, Nith, etc.

What to do?
Do a lot of research; consult with any local experts. I'm convinced this species could be found in all the Borders counties, from the North Sea to the Solway.

Field Notes.

INTERLUDE:
FRESHWATER PEARL MUSSEL (M. margaritifera)

This freshwater invertebrate may live for 130 years and has been an important part of Scottish lore for hundreds of years. Although it is found on several continents, its history is very closely allied to Scotland, particularly the Highlands. This creature can produce valuable and beautiful pearls, which have been collected for centuries. Many members of Scotland's travelling community were associated with the pearl trade, particularly in the Highlands and North East. It was estimated that over 160 rivers, streams and lochs once had freshwater mussel populations. However, recent surveys have also shown that this mollusc has become extinct in many places and rare in many more. Extensive surveys are proving that its range has shrunk alarmingly, both in Britain and the rest of its world range. The species is now protected against any fishing or interference but illegal poaching still goes on, which is now classified as wildlife crime. A good citizen and naturalist should report any illegal activity. The species is both an indicator of water purity and its siphoning properties help insure water quality where the mussel is still found so it is an important part of our biodiversity.

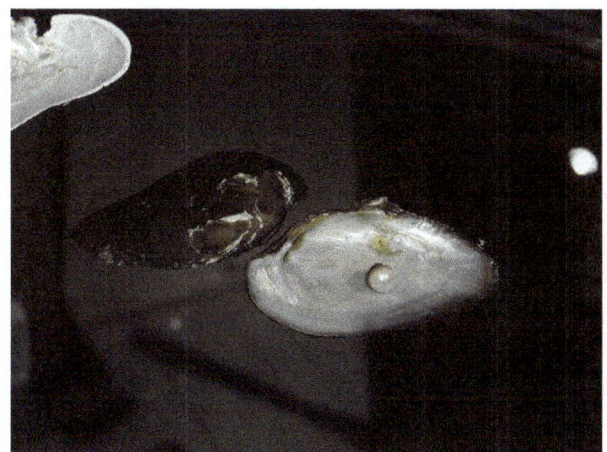

The pearl mussel has been part of British history for thousands of years but only gained full protection in 1998.

Its role in the Scottish Borders is uncertain. It is thought to have existed in streams feeding the upper Tweed and perhaps in smaller, lesser known rivers and streams. Certainly, it was (and is) present in certain parts of Dumfries and Galloway. Today, its habitat is mainly Highland. Although it is illegal to interfere with the species, if examples of this rare animal are found please report any findings to the appropriate wildlife authorities.

Mystery Animals of the Scottish Borders

FISH
From the Solway to the North Sea, The Scottish Borders is served by river systems famous in legend and lore and also offering some of the best trout and salmon angling in Europe: Cree, Fleet, Nith and Urr the Yarrow, the Ettrick, the Teviot, the Tweed, the Eye and many other equally famous streams. The junction pool at Kelso (where the Teviot meets the Tweed) is one of the world's best known angling beats. It is only natural that freshwater offers much to the naturalist and hiker. Naturally, trout and salmon are the most heralded fish but there are many other fish species in local streams that provide some wonder and mystery in the Scottish Borders.

Arctic Charr/Char ('red waimbs')
I prefer the 'charr' spelling and will use it throughout, capitalised.

What is it?
Arctic Charr (*Salvelinus alpinus*)

What is its natural range?
The Charr is a deep-dwelling, schooling fish, native to clear, cold lochs in Scotland and elsewhere in the British Isles. It is a land-locked species. It was once very common in the Scottish Borders, especially in St Mary's Loch, where it became extinct in the 18th century.

Where has it been seen in the Scottish Borders?
Talla and Megget Reservoirs (reintroduced) and waterways in Dumfries and Galloway, including Loch Ken where it probably originated from Loch Doon.

Where to look for it? Clear, deep lochs and connecting waterways where it spawns from November

What to do?
Report any catches to your local angling club, especially within a river system where the fish doesn't normally appear. Charr resemble trout more than salmon and a fish of about 25-30 centimeters would be normal. Males are marked with bright red bellies, especially in spawning times (from late November) The fish are good to eat and usually indicate clear and unpolluted water. They are the

natural prey of ferox trout.

Although the Arctic Charr (*Salvelinus alpinus*) is abundant in my native Canada, I first got to know this salmonid fish well in the Highlands of Scotland, where I often caught several in a deep waterfall pool between Cam Loch and Loch Veyatie in Sutherland. I never caught the fish on the fly because it is not generally a surface feeder. Unlike similar species (trout and salmon) it is not solitary and territorial but is a shoaling, communal fish. Unlike the Canadian Charr which is sea-going during its life cycle (hence its much larger size) the British species was cut off by the Ice Age and forced to live its entire life in freshwater. Its presence in any loch or stream also indicates good purity of water. There are many localised names and variations but many Gaelic names refer to its red belly: red trout, *gillaroo* (red boy) It can still be caught in many Highland Lochs: Veyatie, Cam, Loch Ness (and many more) but its range is restricted by pollution and other factors. It is also a favoured food fish of the enormous trout known as Ferox, gigantic brown trout that regularly feed on other fish. Where there are big Ferox trout there are usually schools of Charr. Charr are generally smaller than trout and a fish of one foot or one pound is a good catch. The record Charr in Scottish waters was caught in Loch Arkaig in 1995; it weighed 9 lbs 8 oz. "The Fish Species of the Tweed and Eye Catchments" also gives some Border background to the charr:

This study quotes a description of St. Mary's Loch in the Borders, made in 1649. Words within parentheses are my own:

"There is also taken in thir lochs a little fish, called by the country people Redwaimbs (red wombs or red bellies) It is about the bigness of a herring and the belly of it wholly red. It is never seen nor taken but between All Hallows (1st November) and Martinmas (11th November) the space of 10 days, and that only in the little stream that runs betwixt the two lochs. It is very savoury meat; I rank the charr as tastier than trout, sea trout or even salmon, perhaps owing to water depth and purity and their diet) and at that season the country people, with plaids sewn together like a net, have taken such store of them that they carried them home and salted them up in vessels for the food of their families."

By 1722, no mention is made of them locally and it is supposed that they had become extinct in St. Mary's Loch during that century. My own field work with Highland Arctic Charr has shown that like other salmonids, they breed in cold late autumn into winter but that their spawning habits probably differ greatly from loch to loch, depending on local conditions. The St. Mary's Loch description sounds accurate to me. However, such massive capture of breeding fish for many decades probably altered both spawning patterns and numbers.

Hallowell also records that within the Borders river catchment area:

"The species was re-introduced to the catchment, to the Meggat and Talla Reservoirs, from Loch Doon in Ayrshire by Scottish Natural Trust in the early 1990's...It is just possible that there was or is a second population of charr within the Tweed catchment. The account of the Parish of Roberton for the <u>Old Statistical Account</u> (Sinclair 1815) that was written in 1792/93 mentions that there were "red trout, much resembling that of Lochleven" in one of the lochs-not Alemuir-in the upper catchment of the Ale. The only fish from Lochleven likely to be called a "red trout" would be Charr..."

Dr. Peter S. Maitland of the Fish Conservation Centre in Stirling wrote an excellent study of the Arctic Charr in the south of Scotland which includes much of the historical Scottish Borders:

THE STATUS OF ARCTIC CHARR, SALVELINUS ALPINUS (L), IN SOUTHERN SCOTLAND, A CAUSE FOR CONCERN

Maitland notes that all disappearances of Charr populations have occurred in the south of Scotland. As noted previously, St. Mary's Loch in the Borders was had a thriving Charr population. Maitland quotes an earlier historical source: "In this Loch are Trouts, Eels, Pearches, Pikes and a kind of Fishes called by the Country people Red Waimbs from the bloud red colour of their Belly. The fish itself is about the bigness of a herring with a forked tail. The Herds about Michaelmass used to take great number of them catching them in their Blankets at a little Rivulet that comes from the Loch of Lowes into this, the two Lochs being almost one and divided by a very small nick of ground..." *The Statistical Account of Scotland for 1794* does not mention Charr in St. Mary's Loch and more recent echo soundings and gill-netting of the Loch recorded no Charr.

Another Borders Loch, Loch Grannoch in Galloway, also had a native Charr population, mentioned by the traveller Thomas Pennant in 1769. They were called "red-wames". These fish were recorded well into the 19th century but by the 1950's had long been gone from the Loch, despite

FIG. 1. An arctic charr from Loch Doon, Ayrshire.

attempts to net specimens. Recent gill-netting operations produced no Charr and revealed a much depleted native Brown Trout population.

Loch Dungeon in Dumfries and Galloway had a thriving Charr population as late as 1900: Maitland quotes Moss, writing for The Dumfries Courier in 1899: "Shortly since I had the pleasure of receiving a fine specimen of Char from a friend, who had caught it in Loch Dungeon with a fly when angling for trout. A few are caught each year in this way in the loch..." The last specimen reported was in 1952 and it is believed there are no Charr left in this loch.

Maitland also notes that Charr reported in the Ken river system in Dumfries and Galloway are assumed to have come from Loch Doon in Ayrshire, where they still exist. Rare specimens crop up in Loch Ken itself and are also assumed to come down from Loch Doon via waters used by the hydro-electric scheme.

Maitland notes the effects of acid rain (acidification) afforestation, engineering schemes, angling and fish farming. Reintroduction of Charr has successfully been undertaken in at least two Borders reservoirs: Talla and Megget, which is also linked to St. Mary's Loch where Charr became extinct.

My own experience is that Charr prefer deep water hence live normally in lakes/lochs but ascend streams and burns to spawn, usually from November onwards. Because of depths Charr thrive in, anglers are more likely to take them on wet fly rather than dry and are probably most likely to take them on artificial lures whilst perhaps spinning or trolling for pike, ferox or other fish. I would encourage anglers in the Scottish Borders to be aware of this brightly-coloured relative of trout and salmon and note the details of where and how the fish is caught. Its role as a barometer of clean water is important to all naturalists and anglers. Its mystery lies in how it became extinct and where and how it still thrives. The Charr in full spawning colours is a natural treasure. Many fish species are prevented from expanding by natural features such as the sea, mountain ranges, deserts or man-made barriers such as cities and motorways. However, any reflection on rare or unknown species must wonder if a rare or extinct species was last seen (or caught) only a few river systems away, it might be possible that eggs or hatchlings might been transported by birds, anglers or other agencies. I give you the curious case of the **Burbot** (Lota lota) one of the British Isles more recent extinctions. The last Burbot was reputedly caught in the River Cam in Cambridgeshire in 1969, having once had a wide range in English rivers ranging from County Durham to Cambridge and maybe further south. It was noted in the Trent, the Ouse, the Cam and other rivers. It had many local names such as the *Blob-kite,* the *Burbolt,* the *Cony-fish,* the *Eel pout.* Generally, they were thought to be good eating. A good size Burbot might weigh five pounds and be two feet in length. Given its range right up to the Scottish Borders, it seems some of its preferred habitat might have been found in Dumfries, Galloway or Wigtownshire rather than the eastern Borders, including the Tweed catchment, where it has never been recorded. The

Burbot was mainly a northern cold water fish yet I have yet to find any Scottish references to this curious fish, once favoured throughout the British Isles. It is best to think of this as a long, freshwater codfish, with a barbel (like a catfish) It is still common in northern climes and is caught in traps in my native Canada where it is a desirable food fish, especially amongst First Nations peoples. I can vouch for its excellent flavour. Scottish naturalists or anglers, if you ever come across something resembling a freshwater codfish, you may have proof that the fish didn't become extinct in English or Scottish waters. Indeed, in 2010, anglers reported catching the fish across the border in Cumbria, but this catch was never verified.

Border anglers have had a long history of landing record catches, including the Common Eel (*Anguilla anguilla*) a record currently held. In *Kelsae*, by local Kelso man and historian Alistair Moffat, a quote indicates how plentiful eels were (and are*)* *"In our summer holidays, we fished for eels. There were a tremendous number in the river then. Every family...at that time kept hens. There wasn't much to feed them on, so they got the eels."* Huge conger eels are also a feature of the sea off the Berwickshire coast. Anyone walking past D.R. Collin, the Eyemouth fishmongers in Kelso, was struck by a photograph in the window of a record Conger eel, whose weight and length needed support by a team of men.

In *The Fish Species of the Tweed and Eye Catchments* other species mentioned include the familiar Sea Trout, Smelt/Sparling, Flounder, Lampreys (three species) Baggie (minnow) Stone Loach, Roach, Gudgeon, Dace and Bullhead. Some of these are historical instances and some clearly introductions as bait use for other fish. It is not clear how many of these species have vanished completely or breed currently in the Borders water systems. Most of the angling clubs require records of catches, mostly brown trout and salmon but other species can be monitored and noted.

The Ferox
What is it? Ferox (*Salmo ferox* or *Salmo trutta*)

What is its natural range?
There is scientific debate about whether or not the Ferox is a separate species (Salmo ferox) or a variety of brown trout (Salmo trutta) but with different life habits, diet and growth patterns than other brown trout.

Where has it been seen in the Scottish Borders?
It has been widely reported: St. Mary's Loch, the Upper Tweed, the Upper Annan and the Upper Deveron. It is also stocked by angling clubs.

Where to look for it?
It prefers deep lochs and is often found with its favourite prey item, the Arctic Charr.

What to do?
Report any catches to your local Angling Club or Association.

There is much mystery and controversy surrounding this giant brown trout. Is it a separate species or just a variation of lifestyle of the brown trout? The Ferox grows much bigger and is fond of eating other fish, particularly smaller brown trout and its favourite prey: the Arctic Charr. I can offer a few examples. Many years ago, while walking the shores of Loch Veyatie in Sutherland, I saw a dead fish lying on the bank. At first, I didn't recognise it. It looked like a brown trout but was elongated, with a pike-like head and kype (long lower jaw) Moreover, I guessed its weight at five to six pounds, much larger than any brown trout I'd ever caught or seen. I'd seen my first Ferox. Veyatie was also a great Charr loch, the favourite food of the Ferox. In July 1938, in the same loch, the Reverend Walter E. Lee caught a monster trout (Ferox?) weighing 16 lbs, 32 inches long, with a 26 inch girth. (see photograph below)

This monster trout is found in many lochs in the Borders; most significantly it is still present in St. Mary's Loch, once plentiful with its food source, the Charr. Deep glacial lakes are its preferred habitat. It is worth having some background on this mysterious fish. The name itself was coined in 1835 by angler and scientist Sir William Jardine. R.N. Campbell wrote in "Journal of Fish Biology 14, 1979":

"Ferox are long-lived, late-maturing, piscivorous brown trout which in Britain and Ireland, are often present in large, deep glacier-formed lakes containing Arctic Charr..."

The main argument is whether Ferox is a separate species (*Salmo ferox*) or just a brown trout differing in habit and life history from the wild brown trout (*Salmo trutta*) Studies have shown that Ferox from different lochs are more genetically similar to each other than they are to brown trout in the same loch. The jury is still out on this and much more study is needed on why brown trout opt for the Ferox 'lifestyle.'

From the **TweedBioBlog 13 May 2014** from 'Ferox and Fungus'

"Apart from general background, our particular interest in these fish is that there might be river populations of ferox or fish with some ferox ancestry and populations of long-lived an unusually large trout are now known from the upper Tweed, upper Annan and upper Deveron. These have not been genetically examined yet...However, the large Sea-trout of the Tweed and Northumberland rivers are unlikely to be ferox type as they are not long-lived even though they can get much larger than the Waterville fish (large sea-going ferox in Southwest Ireland)

The mystery continues:
Do we have Ferox sea trout in the Borders?

Just how big do these fish get?
Given that Charr have become rare in St.Mary's Loch, how did the Ferox population survive there without their favourite prey?

Why do some wild brown trout become Ferox? (there have been three anglers in Scotland who have caught Ferox exceeding 30 lbs. Moreover, the current angling record weighs in at 31 lb. 12 oz.) Using the 40% giantism rule (more on this later) that suggests monsters of over 40 lbs might exist in Borders lochs!

How does the Ferox relate to another borders predator, the pike?

In 2020 Carl Marshall referencing *Short Sketches of the Wild Sports and Natural History of the Highlands* by Charles William George St. John. Publisher: John Murray, London; First Edition (1846) wrote an article about ferox trout and possible links to the Loch Ness Monster for the CFZ Journal *Animals & Men*. It is reproduced as an appendix at the back of this book.

The great Loch Ken pike
Any sightings of mysterious creatures must include possibilities of giantism, that the animal we are seeing is a larger version of what we normally expect. One of the Borders most celebrated stories is surely that of John Murray's 'monster pike.' John Murray was a gamekeeper to Lord Kenmure. Murray died at Kenmure (Galloway) on 3 January 1777. His

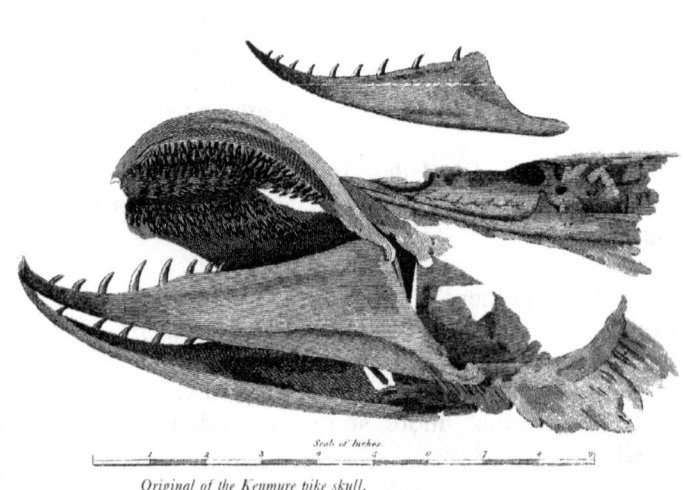

Original of the Kenmure pike skull.

gravestone inscription says it all: "To the memory of John Murray, who died at Kenmure Jan. 3, 1777". Inscribed below are a gun, fishing rod, dog and partridge.

Given to lively talk, snuff and practical jokes, Murray caught his monster pike in Loch Ken. It weighed 72 pounds and measured over 7 feet in length! Murray was said to have carried the pike over his shoulder, the huge tail dragging the ground. Throwing the pike at Kenmure's feet, Murray quipped: "Ye may catch the next yersel..."The skull of this huge fish was said to be preserved for many years at Kenmure. By comparison, the largest pike recognised as a Scottish record was caught in 1947, weighing nearly 48 pounds.

Pike have also featured in the eastern waters of the Scottish Borders. *The Fish Species of the Tweed and Eye Catchments* mentions pike as probably being brought to the Borders by monks in the Middle Ages as food fish in the 12th century. It was caught by Tweed anglers more often in the past. It is still found in St. Mary's Loch but is less common as a river fish. Anglers were offered bounties for pike in order to protect salmon and trout.

Grayling. *Thymallus thymallus*. This salmonid fish is now common in Borders Rivers, particularly the Tweed and Teviot. It was probably introduced to the Teviot from a pond at Monteviot from the south of England. This delicate-mouthed fish offers great sport on the fly in the salmon and trout closed season. Its scientific names suggests a flavour of thyme. This fish is not common throughout Britain so does fill a niche both in nature and in the angling calendar.

It is a handsome fish, most recognisable by its distinctive dorsal fin. In the 1870's it was plentiful enough to be sold in the fish shops of Berwick, supplied by netters on the Lower Till. However, many Borders anglers have never seen or caught a Grayling, adding some small mystery about the fish and its habits and whereabouts.

The Vendace (*Coregonus vandesius/Coregonus albula*)

What is it?
Vendace

Lochmaben, Dumfries, four miles west of Lockerbie and eight miles from Dumfries itself, boasts three lochs: Kirk Loch, Castle Loch and Mill Loch. The latter two lochs were also unique in Scotland and the British Isles as being an indigenous site for the endangered and "the rarest freshwater fish in Britain." The Vendace (*Coregonus vandesius/Coregonus albula*) It is a small fish, 20 cm/8 inches or so, but very tasty and much prized as a delicacy.

Mystery Animals of the Scottish Borders

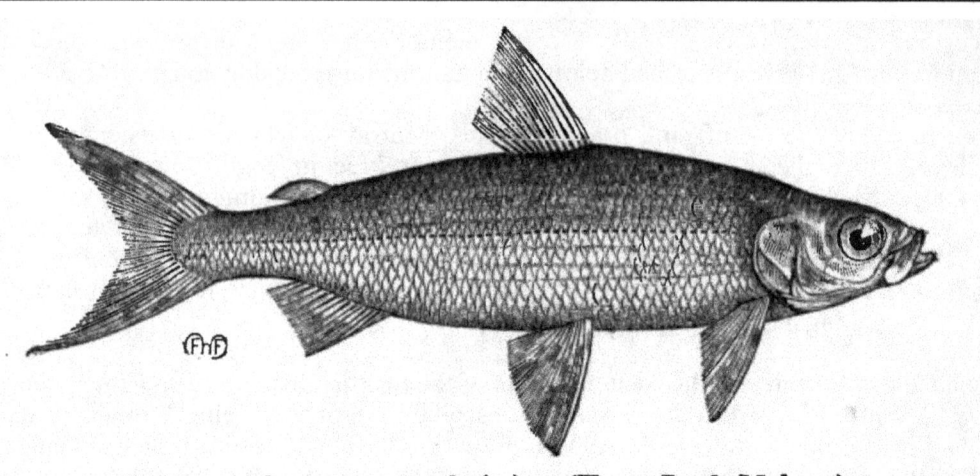

Vendace (*Coregonus vandesius*). (From Loch Maben.)

What is its natural range?
The Vendace was once more common in deep clear lochs/lakes in the British Isles. This small white fish was known as a native species at four sites in Britain: Bassenthwaite Lake (last recorded there circa 2001)and Derwent Water in the Lake District and Castle Loch and Mill Loch in Lochmaben. It only survived in Derwent Water, disappearing from Castle Loch (circa 1910) and Mill Loch (circa 1990's)

Where has it been seen in the Scottish Borders?
It disappeared from the lochs around Lochmaben: Castle Loch, circa 1910 and Mill Loch circa 1990's. It has been found nowhere else in the Scottish Borders.

Where to look for it.
Although the native populations vanished in the Scottish Borders, in the mid-nineties fish from Bassenthwaite and Derwentwater were translocated to two sites in the Scottish Borders, Daer Reservoir near Biggar in South Lanarkshire and Loch Skeen in Dumfries and Galloway. Although there has been some doubt about the success at Daer Reservoir, the Loch Skeen population appears to be thriving.

What to do.
Contact your nearest Angling Club or Association. It would be extremely rare to find it outside Daer Reservoir or Loch Skeen.

Yarrell gave some historical background to this rare fish: "The Vandace is well known, to almost every person in the neighbourhood (Lochmaben) and if among the lower classes, fish should at any time form the subject of conversation, the Vandace is immediately mentioned, and the Loch (Castle Loch) regarded with pride as possessing something of great curiosity to visitors, and which is thought not elsewhere to exist...An idea prevails that this fish, if once taken from the water, will die, and that an immediate return will be of no avail; and it is also believed that it will not exist in any other water except that of the Castle Loch. The fish is of extreme delicacy ..." The writer goes on to describe it as a shoaling fish living at great depths and taken with nets rather than line. They appear to spawn in November onwards (like the Charr) He concludes: "During the summer fishing parties are frequent...a club consisting of between twenty and thirty of the neighbouring gentry possessing private net, meet annually in July to enjoy the sport of fishing and feasting upon this luxury." Although the native populations vanished in the Scottish Borders, in the mid-1990's fish from Bassenthwaite and Derwentwater were translocated to two sites in the Scottish Borders, Daer Reservoir near Biggar in South Lanarkshire and Loch Skeen in Dumfries and Galloway. Although there has been some doubt about success at Daer Reservoir, the Loch Skeen population appears to be thriving. Like Charr, the Vendace is a barometer of clear and deep water and indicates that water conditions have improved generally within its historical range. The fish is both protected and endangered. Any anglers should note/photograph their catch and release it immediately and safely. The Vendace illustrates a twofold mystery: how such a lovely fish could become extinct and how it could also thrive once re-introduced.

Scottish anglers have long been spoiled by having such productive trout and salmon streams but other fish also provide much sport and enjoyment to the angler. The Scottish Federation for Coarse Angling maintains records for various fish and many listings are in the Scottish Borders catchment. Here are a few recent examples: The largest Chub was caught in the River Annan. The largest eel was caught in 'an undisclosed Border Quarry' in 2007. Another salmonid species, a record Grayling (*Thymalus thymalus*) was caught in the River Tweed in 1994. The record Tench was caught in Castle Loch, Lochmaben in 2003. Seeing my later notes in this book on giantism, it is pertinent that large fish ('giants') also inspire some mystery and wonder, posing the question if even larger ones exist, particularly if the record has lasted some decades.

Atlantic Salmon (*Salmo salar*) The Leaper, the King of Fishes. The salmon has

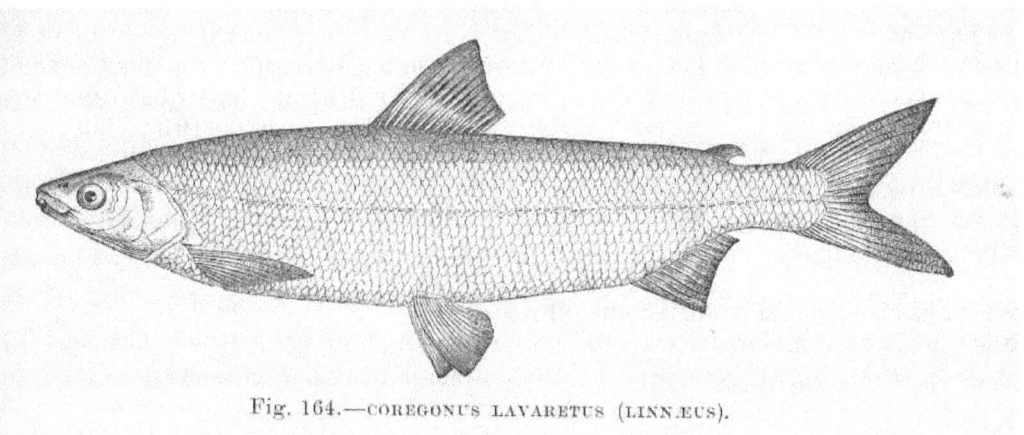

Fig. 164.—COREGONUS LAVARETUS (LINNÆUS).

Mystery Animals of the Scottish Borders

been an important fish in the Scottish Borders since Neolithic times and it is easily the most prized fish in local river systems. Georgina Ballantine caught the biggest recorded salmon on 7th October 1922 in the River Tay. A fibreglass cast of that monster can be seen in Perth Museum. It weighed 64 pounds!

However, such monstrous fish were not unknown in the Borders. In Yarrell (William Yarrell, *A History of British Fishes*.1841) the then Earl of Hume is quoted:

"My uncle, my father's elder brother, caught a Salmon (in the Tweed) with a rod which weighed sixty-nine pounds and three quarters." A photo of the monstrous Ballantine fish dwarves one of the angler's nieces.

What is it? Burbot (*Lota lota*)

The last burbot was reputedly caught in the River Cam in Cambridgeshire in 1969, having once had a wide range in English rivers ranging from County Durham to Cambridge and maybe further south. It was noted in the Trent, the Ouse, the Cam and others. The Burbot was mainly a northern cold water fish yet I have yet to find any Scottish references to this curious fish, once favoured throughout the British Isles.

What is its natural range?
It was found in slow-moving rivers on the east coast of England. It has never been reported in Scotland.

Where has it been seen in the Scottish Borders?
It has never been reported anywhere in Scotland yet as late as 2010, anglers reported catching one across the border in Cumbria, further north than ever reported.

Where to look for it?
This would be quite a discovery since the fish has been thought to be extinct in Britain since 1969.

What to do?
Report it to your nearest angling club or association. Inform local wildlife agencies.

Field notes

Mystery Animals of the Scottish Borders

AMPHIBIANS
One of Scotland's rarest and unique amphibians is found only in an isolated part of the Scottish Borders. It is the Natterjack Toad (*Bufo calamita*) which takes its name from jack=toad and natter=they are noisy and have a unique mating call, the loudest of any European amphibian. Although they are more common on the Continent, they are limited in Britain to a few locations in England and only one location in Scotland, the sand dunes and salt marshes along the north coast of the Solway Firth, in Dumfries and Galloway. More particularly, they are most concentrated around he Caerlaverock National Nature Reserve. They hunt and prosper in areas of short grass and heathland. They need fresh water (or slightly salt) to breed. They are nocturnal and feed on a variety of insects and invertebrates. They are also eaten by birds and foxes, hedgehogs and otters. Unlike the only other toad in Britain (The Common Toad) Natterjacks are able to run rather than hop and are identified with a telltale yellow stripe on the back. Conservation of the toads in Dumfries and Galloway is a real success story, with an increase of nearly 400% of breeding males in three years due to an awareness (and protection) of their breeding habitat. This toad's range was once much wider but it is still unlikely to be found outside of the Caerlaverock area.

Amphibian (toad)
What is it? Natterjack Toad (*Bufo calamita*)

What is its natural range?
They are found on sand dunes and saltmarshes in very limited habitats in Continental Europe and Britain. In England, they are confined to East Anglia and southern England and a few habitats in northwest England. It has been called "the running toad" because unlike other toads, it appears to run rather than hop.

Where has it been seen in the Scottish Borders?
They are only found along the North Coast of the Solway Firth in Dumfries and Galloway but are most abundant around the Caerlaverock National Nature Reserve in Dumfries and Galloway.

Where to look for it?
In the habitats/locations described above. They are nocturnal. The species only occurs in sandy soil. They hibernate in winter. They are smaller than the only other toad found in Britain (The Common Toad) The Natterjack has a distinctive yellow stripe down the centre of its back. Their short legs means they can run rather than hop like other toads or frogs.

What to do?
These unique amphibians are rare and protected and are very unlikely to be seen outside their known range. Report any sightings (back up with photographic evidence and other details) Do not disturb these creatures. Report to: Scottish Natural Heritage, The Wildfowl and Wetlands Trust and the Scottish Wildlife Trust (addresses in bibliography)

REPTILES

Reptile (snake)

What is it? Adder (*Vipera berus*)

What is its natural range?
It is found throughout Scotland and is our most common snake. It is venomous. It prefers heathland, moorland, scrub and sand dunes. They feed on small mammals, including mice and voles but they also eat amphibians such as frogs and toads. They are often seen near water and are excellent swimmers.

Where has it been seen in the Scottish Borders?
It has been sighted in every Borders county. Do not disturb the animal; they are a protected species.

Where to look for it?
Adders are more common than you think but spook easily. They are also well-camouflaged. Find a spot where you think there are adders. Late February or early March is a good time, since the increased daylight and warmth invites them to sun on rocks. Don't move, be quiet and just wait. Use binoculars to scan any large rocks and I almost guarantee you will see a basking adder, usually a male. I have sat on rocks right next to them.

What to do?
Enjoy the moment. Report it at your discretion but remember adders might react to being surprised by a child or pet so it might be best to let walkers know there is an adder nearby if youngsters or pets are present. It is best to just enjoy the sighting then leave the animal to its own peace and quiet. They eat lots of mice

Mystery Animals of the Scottish Borders

and small rats so serve us well. They have a long pedigree in Britain and are part of our heritage.

Reptile (snake)
What is it? Grass Snake (*Natrix natrix*)
This snake is non-venomous and may reach lengths of 180 cm/6 feet making it the longest snake in Great Britain. It's possible that its northward range is being extended due to long-term changes in temperature and it is now more likely to be seen in Scotland. It is relatively common in our two neighbouring English counties, Cumbria and Northumberland so transference over the border is likely.

What is its natural range?
It feeds almost exclusively on amphibians so it will be found where toads and frogs are found. It is often found near water and is a good swimmer. Although it has been common and widely distributed throughout England, it is not present in any but the southern part of Scotland, which includes the Scottish Borders in the widest sense.

Where has it been seen in the Scottish Borders?
It should be remembered that this species can be an escaped pet. However, there have been at least six 'confirmed or possible' sightings in the wild in Dumfries and Galloway

Where to look for it?
Look for it at woodland edges but where there is water nearby, especially where frogs are also abundant.
What to do?
Gather any evidence, get some photos if possible and then report it to the nearest office of Scottish Natural Heritage. There is a lot of study and research being done

into Scottish records for this species. A few decades ago, Scottish sightings were very rare but are increasing. Dumfries and Galloway have had recent confirmed sightings but it is increasingly likely that the Grass Snake may be seen in the eastern Borders as well. It is crucial to brush up on the differences between the Grass Snake (non-venomous) and the Adder (venomous) although it is fairly easy to tell them apart by body shape if nothing else. Although the Grass Snake may legally be kept as a pet, it is a protected species in Scotland and may not be harmed. Under no circumstances should you allow anyone to harm a snake.

Field Notes.

Mystery Animals of the Scottish Borders

INTERLUDE: Beyond mere mystery:
Beasts of Myth, Legend and Nightmare.
Introduction: This book has mainly presented animals that did exist in the Scottish Borders (bears and wolves) and animals that have perhaps become locally extinct, then reintroduced (The Arctic Charr) and finally, real beasts that need identification (the Big Cats) of Cardrona, Glentress, Galloway and other parts of the Borders.

THE LINTON WORM
Possibly the most famous of these legendary Borders beasts is the Linton Worm. The story is told best by Sir Walter Scott in his *Minstrelsy of the Scottish Borders*. This was a collection of Border ballads published in Kelso in 1802, which introduced Borders lore to a greater audience. Two volumes appeared in 1802 and a third followed in 1803.

Linton is a small Roxburghshire community six miles south east of Kelso and a mile from Morebattle. Its name is said to come from "lind": a fiery serpent. It is hilly agricultural country. *Orm* or *Worm* is Norse and Old English for a serpent or dragon and some of its variants are Wermes, Wyrrms, Orms, Stoorworms and Vurms. The Linton Worm parallels the Lambton Worm of Northumberland as well as the more famous legend of St. George and the Dragon. I'll retell the Linton Worm story, culled

from many sources but essentially the same.

Sometime in the 12th century, Linton locals were terrorised by a serpent or dragon, described as "in length three Scots yards (note: A Scot's yard was a textile measure of about 37 inches and equates to an "ell". The English Yard was 45 inches) and bigger than an ordinary man's leg-in form and callour (colour) to our common muir edders (adders)." The beast lived in a lair (still known as the "Worm's Den") in the hollow of a hill. At dusk it would emerge to eat crops, livestock and even people. Weapons such as arrows, huge darts and spears were useless against it. However, William (some say John) de Somerville, the Laird of Lariston devised a plan to rid the country of this loathsome Orm. He studied the habits of the animal (from a safe distance!) and observed that the animal would often lie motionless with its huge mouth agape. He had a local blacksmith forge a special barbed spear. On it, he placed some burning peats dipped in tar and brimstone. Accompanied by a "stoute servant" de Somerville found a good time to approach the beast at the mouth of its lair and plunged the spear and peat into the creature's mouth. The beast writhed and thrashed, its death throes forming the contorted local landscape now known as "wormington". For his bravery, Somerville was made Royal Falconer to King William, knighted and made "First Barrone" of Linton. The Somerville crest appropriately features a Wyvern (Dragon) A carved stone at Linton Kirk honours his bravery.

These Isles seemed awash with giant land serpents at this time and the stories of these creatures are remarkably similar. The unique charm of the Linton Worm legend is that it is very specific as to the beast's size and colour. An adder of "three Scot's yards" (111 inches, nearly 10 feet long) would have been remarkable if not impossible but its fiery breath and hypnotic eyes and death thrashings forming a whole landscape are clearly fictional add-ons, maybe to enhance the reputation of its slayer. Using farm tools as weapons, your average Borders farm labourer was more than capable of dispatching a big adder; nor would he (or she) require a laird on horseback to do it for him. Adders, though secretive and wary, are still not uncommon in this part of the Scottish Borders. I have seen them on sunny February days, soaking up some rare heat. They are sluggish and non-threatening. The real mystery though is what inspires these stories and how much truth is in them? We can only imagine a Borders landscape with bears, wolves, lynx and other vanished creatures. Perhaps the notorious Orm is simply a marvellous but extinct animal: not a dragon but a huge reptile maybe flourishing in a British climate very different from today's? After all, the imaginary Linton serpent was much smaller than a very real anaconda, Komodo Dragon or salt water crocodile.

Just as the Linton Worm is overshadowed by its English counterparts (The Lambton Worm and St. George and the Dragon) so does the Linton Worm eclipse another Borders beast. The story is told in Alan Temperley's excellent *Tales of Galloway*, a perennial source of Borders lore. Go to Temperley's book for *The White Snake* in its entirety but I'll humbly summarise the tale here:

In the mote of Dalry on the east Bank of the River Ken a giant white snake (serpent) took possession of an overlook. As a tribute to the fearsome creature, locals fed it all their dairy milk as well as the occasional sheep or cattle. If the beast was not satisfied with their offerings, it would simply go into the community and eat whatever it pleased, villagers included. The serpent grew and grew and was soon able to save time by swallowing everything whole, without chewing. The local blacksmith named Michael Fleming had enough when the creature devoured the corpse of his buried wife. The clever smith fashioned a special suit of armour with protruding sharp knives that could spring into action like porcupine quills. He carried two long knives as weapons. The snake grabbed him and swallowed him whole! However, he was still conscious and able to slash at the insides of the snake. After much thrashing, the snake died with the smith still inside! Fleming slashed his way out. The blacksmith was hailed as a hero and carried through the town in triumph. Later, they cut the snake up into bits and floated it down the river where according to local lore, the snake's are still buried in the mud of Loch Ken.

To complete a hat trick of such beasts, the Solway Firth (forming the border between Cumbria and Galloway) has a beast which is both comfortable on land and in the sea. The Sea Worm fed voraciously on fish and cattle. This dragon met an untimely end when it was stranded at low tide and became impaled on wooden stakes.

INTERLUDE: GIANTISM
This topic follows on naturally from the Linton Worm which could have been a truly gigantic reptile. When looking for 'mystery animals', any field observer needs to be aware that an abnormally large example of a known species might appear to be an entirely different animal, especially when dealing with invertebrates like insects or arachnids. There is a logic behind it: "That can't be species x because we have never seen such a big example of species x." One argument against the existence of giant squid in the 1870's was that the squid fishermen in Newfoundland were encountering squid that were so much bigger than the smaller edible examples already known. It

didn't help that the species **Architeuthis** was called a giant squid when in truth, architeuthis was simply a big squid, and a totally separate species from smaller known squids. I can give a more personal example. As youngsters, we fished in all the local rivers and ponds. One day, we were fishing in the West Fork of White River, in Morgan County, Indiana in the American Midwest. Using live bait, we were after panfish (bluegills/sunfish/crappies) and small catfish. Invariably, we caught smaller river turtles, including the Common Snapping Turtle (*Chelydra serpentina*) These latter weighed a few pounds and had shells roughly the size of a dinner plate. However, using hand lines, we were hooking something much bigger. The shell was about the size of a small table. This turtle itself was perhaps five times the size of the turtles we normally caught. One day, we managed to get enough of the turtle out of the river for a better look, before it broke our hand line and disappeared back into the current. Nobody believed the size of turtle we described. However, many years letter, the local wildlife department verified that the Alligator Snapping Turtle (*Macroclemys temmincki*) a separate southern species several times the size of the Common Snapping Turtle, had been found in the local river system. It's likely we were the first to make that discovery, but being boys, weren't believed. However, it's useful to have a formula that might give a naturalist a reasonable measure of what a giant of any particular species might look like or measure. Enter Bernard Heuvelmans ((born in Le Havre in 1916-died 2001) Anyone who searches for unknown animals knows the name of this man who is easily the 'Father of Cryptozoology': the search for unknown or undiscovered (or mysterious) animals. He wrote two books that are the likely two books to be stranded with on a mythical desert island: *In the Wake of Sea Serpents* and *On the Track of Unknown Animals*. On page 306 of my English edition of *In the Wake of Sea Serpents* Heuvelman's leaves us with a working formula: He quotes Dr. Maurice Burton, formerly of the British Museum. If we have trustworthy measurements for the species in question "we find that the outsizes of species are: 68 per cent above average size, and 40 per cent above the unusually large." Heuvelmans goes on to say that if a large sea snake is normally 10 feet long then a giant of that species (40 per cent) would be 14 feet long. Heuvelmans uses this measure (rightly I think) to show that whatever sea serpents are or aren't, the descriptions of the largest ones are much longer than a giant sea snake. This "40 per cent above the unusually large" proves a handy rule of thumb, useful for fishes, reptiles and other creatures often described by length rather than weight, including descriptions of big cats. It also might apply to invertebrates. I shall rely on this formula elsewhere in this book.

INTERLUDE: THE GIANT SQUID
This topic naturally follows Giantism.

Below the thunders of the upper deep;
Far, far beneath in the abysmal sea,
His Ancient, dreamless, unrivalled sleep
The Kraken sleepeth:

(Tennyson *"The Kraken"* 1830)

When looking for mysterious animals, I would like to coin a term: the 'anti-unicorn effect'. Over time and with the advancement of science, instead of a presumed 'real' animal like the unicorn later becoming myth or superstition, a supposed mythical animal becomes very

Fig. 1. Architheutus dux
(Expl. i Bergens Museum) ca. 1/25 nat. st.

real. In ancient days it may have been called a 'sea troll' or 'Kraken' but the Giant Squid is one good example of this effect.

Mariners (particularly whalers) had long reported a huge sea creature with many arms which was sometimes longer than their boat or ship! Gigantic squid began to appear off Newfoundland in the 1870's and some of the encounters were harrowing. For a full account of this, I would absolutely recommend Bernard Heuvelmans' *In the Wake of the Sea-Serpents* (my edition is the translation from the French published by Rupert Hart-Davis, London, 1968) Anyone interested in the history of the giant squid must begin with his Chapter Two: "The Kraken and the Giant Squid" Earlier than Newfoundland was the account given by the French gunboat Alecton (see previous illustration) which brought this creature to the attention of a wider (and sceptical) scientific world.

On the 30th of November, 1861 the Alecton encountered a huge 'floating octopus' in the sea off Tenerife. An octopus has eight tentacles (hence its name) but the squid has ten but that fact wasn't generally known then. The ship pursued the creature and fired bullets (to no effect) then harpooned the creature. When hoisting it on board, its tail broke off and that was the only evidence the French gunboat retained. The total length of the creature was over thirty feet. However, the account was later met with scepticism by the French scientific community and the evidence was rejected. However, by the end of that century the existence of the creature had earned it a scientific name: Architeuthis

We now know that these squid are not monstrous anomalies of a small species of squid but simply a genus and species of very big squid: *Archteuthis dux*: (chief/leader squid)

In more recent years, we have also learned of the Colossal Squid (*Mesonychoteuthis hamiltoni*) native to waters around New Zealand. It is even longer and heavier than *Architeuthis dux.*

What does this have to do with the Scottish Borders? Nothing, yet. I have not been able to find any records of moribund Architeuthis washed up on the coasts of either the North Sea (the eastern Borders) or the Solway coast (western Borders) However, there are countless examples of the Giant Squid in Scottish waters, especially in the North Sea and I infer that such creatures have washed up (and will wash up) on coasts of St. Abb's, Coldingham and Eyemouth.

Although sightings of squid seem less common on our western coasts, there have also been sightings and reports.

Some North Sea records include (from Heuvelmans and other sources) the following; these animals were of various sizes and in various states of decay.

- Shetland 1860-61
- Skateraw, East Lothian 1917
- Caithness 1921/22
- Gullane, East Lothian 1930
- Arbroath 1937
- Scarborough 1938
- Whalefirth Voe, Shetland, 2nd October 1949.

This Shetland specimen was the largest ever washed up in the British Isles. It was over 24 feet long. It was cut up for bait but was identified by jaws saved by Dr. Peterson of Camb. It was reported in The Shetland News on Thursday 20th October:
"On Sunday, the 9th of October a large octopus (sic) was found on the beach at the head of Whalefirth Voe. It had a rounded bag-like body, with a large head bearing eight long arms or tentacles studded with suckers. The longest arms were somewhere around 20 feet, the others six to eight feet. It had two large eyes and a horny beak, resembling a parrot's beak. Its chief food is crustaceans, yet could it be the herring fry, so plentiful in Whalefirth Voe just now, induced it to come so far inshore?"

Here are a few other records of moribund specimens found in Scotland:

- Aberdeen, 1949
- Nigg Bay 1949
- Angus, 1951
- Orkney and Shetland 1956/1957

A Scottish trawlerman, James Slater of the Viking Prestige, has the distinction o capturing a huge squid in his nets on the night of 1st February 1957, off Rattray Head. it was nearly 25 feet in length.

- 1977 North Berwickshire
- 1984 near Aberdeen
-

There are many more unreported or unverified examples of the remains of giant squid on our shores.

The largest examples of these animals are found in the very deepest oceans, much deeper than anything in British coastal waters. However, any walkers in the Scottish Borders along the Solway or North Sea should be prepared for the possibility of seeing a moribund example of this creature once thought impossible to science. This is not likely but not impossible either. Take good photographs and perhaps safely obtain a small sample of tissue for future research because many of these carcasses

are soon scavenged or destroyed by storms or reclaimed by the sea. Report any such carcase immediately to a relevant local wildlife agency. Make your discovery known to local media sources, including online ones.

BIRDS
Many excellent wildlife surveys have been undertaken by the Scottish Ornithologist's Club. Borders Branch. For example, see their Borders Bird Report. No 27. 2011. Edited by Ray Murray.

For example an end of year report noted that in 2010:

Rare: Bittern (seen on the Eye Water)
Hoopoe. Crane
Rough-legged buzzard.

In 2010, Golden Eagle. "only 2 reliable observations"

Much of the most valuable ornithological work is done by the Borders Scottish Ornithological Club (a branch of the Scottish Ornithological Club) Their work consists of recording, discussion groups and most valuable field work. They have also produced a free full-colour guide to birdwatching in the area which can be downloaded from their website. They also direct the serious birdwatcher to *Where to Watch Birds in Scotland*, by Mike Madders and Julia Welstead, which includes many of the best sites in the eastern Borders.

Find out more about SOC Borders branch
https://www.the-soc.org.uk/local-branches/borders

Visit the Borders recording area page
https://www.the-soc.org.uk/bird-recording/local-recorders-network/areas/borders

Where to birdwatch in the Scottish borders
https://www.birdguides.com/sites/europe/britain-ireland/britain/scotland/borders/

Birds in the Borders (a telling anecdote)
In Scotland "where do you stay" means where are you living currently, i.e. at what address or in what town or village. I stay in Kelso where many gardens offer bird feeders and wild fare like apples, damsons and rowan berries. It is not uncommon to see dozens of different species in one garden in one day. Our garden featured the usual garden suspects: robins, blackbirds, great tits, blue tits, starlings, sparrows, and rarer visitors like the sparrow hawk and the winter fieldfares.

My wife and I as her carer, took great delight in our long living room window looking onto a garden of trees and shrubs, where we placed various bird feeders. We also had feeders attached to the window that let us see the smaller garden birds close up. One sunny mid-morning, on the hanging bird feeder were two distinctive birds we had never seen before. They were about the size of a chaffinch or bluetit but very striking in their markings. As computer users we tried various descriptions. Later, my wife contacted me at work. She was excited to say that she had definite identification based on colour and pattern: they were both zebra finches. Zebra finch? We'd listed goldfinch, chaffinch, greenfinches aplenty but what was/is a zebra finch?

We discovered it is the common native finch of central Australia. It is popular in the UK as a cage bird. Still a good story right? A quick trip to the local pet shop solved the mystery. A local man was cleaning his cages and some of his finches escaped. Although he lived over a mile from us, the birds found their way to our generous bird feeders, where we noticed they were bullied and driven away by the local birds. We don't know their fate. There is a lesson here for all who would discover exotic or unknown species. Don't rule out the obvious explanation (see Occam's Razor) However, it was also a thrill to see something at first wondrous and new in our garden.

Here is a partial list of Borders birds, once extinct but now making a comeback with the help of ornithologists and committed bird enthusiasts.

- **Red Kite.** *'Gled'.* extinct by 1900, reintroduced successfully in Dumfries and Galloway.
- **Osprey.** extinct by 1916 due to persecution. Reintroduced in 1959 at Loch Garten and now 200 breeding pairs in Scotland. Was once commonly seen in the Borders.

Saint Cuthbert and the Osprey
Here's a local tale. In the 7th century, St. Cuthbert and a young disciple were walking along the Teviot. They saw an Osprey with a fish in its talons. Probably with divine assistance the boy was able to approach the bird and take the fish (salmon or trout) and offer it to a famished St. Cuthbert for his tea. The saint requested that the boy return half of the big fish to the bird who provided it. The proverb's emphasis was on sharing

Bird
What is it? Golden Eagle (*Aquila chrysaetos*)

What is its natural range?
Isolated mountain ranges with access to good prey: mammals, birds, fish. It is also a scavenger. In Scotland, it is found mostly in the Highlands. There are thought to about five hundred breeding pairs in Scotland as a whole.

Where has it been seen in the Scottish Borders?
It is thought there are 2-4 breeding pairs in the Borders, including Dumfries and Galloway, where there may be three breeding pairs. A breeding programme is currently being planned by the South of Scotland Golden Eagle Project, funded by Heritage Lottery Fund. The plan is to increase the breeding pairs to 16 in the Scottish Borders, where the bird was once widespread.

Where to look for it?
It has been sighted in the Ettrick and Yarrow Valleys. In Dumfries and Galloway, it has been reported at Craigadam and the Carsphairns. An eagle was badly wounded by gunshot in Dumfries and Galloway in 2012. Earlier an eagle was found poisoned in Peeblesshire in 2007.

What to do?
Report any sightings to your local ornithological club or association, or any known bird experts in your local area. Golden Eagle sightings in the Borders are rare, but possible.

RELICT ANIMALS
Unknown or mystery animals can sometimes be a small surviving (often unexpected) population of a species much more widespread in the past. In would be both exciting and unsettling to stumble on a creature you thought had long been extinct. Encounters with mystery big cats in the Scottish Borders may be an instance of coming across animals that have survived from the past. Sometimes, these animals may be the last of their kind. The woolly mammoth, once widespread in Europe, was relict on Wrangel Island for thousands of years after it was extinct elsewhere in Siberia. Often, relict populations exist on islands or isolated mainland areas. Relict animals can also be 'living fossils', animals thought to exist only in the fossil record. The thylacine of Tasmania survived long after its extinction on mainland Australia. The coelacanth (see the 'Extinct' interlude in this book) is the best known example of a 'living fossil.'

OCKHAM'S/OCCAM'S RAZOR
William of Ockham (1285-1349) was an English philosopher and theologian who gave us the sound principle of preferring the simplest explanation to a complex one, all things being equal.

Introducing Philosophy: A Graphic Guide to the History of Thinking. **Dave Robinson and Judy Groves. Icon Books, 2007.**

In Latin, *Entia non sunt multiplicanda praetor necessitate* (do not multiply things beyond necessity) which was not found in any of his works although he did write *Frustra fit per plura quod potest fiery per pauciora* (It is pointless to do with more things than can be done with fewer)

Thomas Mautner. *The Penguin Dictionary of Philosophy.* London, Penguin. 2005.

In Ben Dupre's *50 philosophy ideas you really need to know. London: Quercus,* 2007 he gives the excellent idea of crop circles, *vis a vis* Occam. If crop circles are either made by hoaxers (which has been done often in the past) or aliens, Occam would say choose hoaxers because there is no supporting evidence for crop circles being done by aliens. Allowing for newer evidence, it is reasonable and prudent meanwhile to explain crop circles as the work of hoaxers. Dupre also notes that a simpler explanation is not necessarily the correct one but is a good default until more evidence is forthcoming. This useful book also mentions the KISS principle, a cornerstone of computing and other technical fields. Simply, KISS stands for "keep it simple, stupid."

In searching for mysterious or unknown animals, Occam's Razor might suggest that likely explanations are more compelling than unlikely ones: i.e. if I see an unknown animal, it is probably more likely that I have been confused by an animal known to others but outside my own experience. Many people claim to see huge black cats in the British countryside. Given that black (melanic/melanistic) examples are only known to jaguars and leopards, then rather than a totally new black cat in the British countryside, it might be more prudent (ala Occam) to think of an escaped leopard or jaguar, or perhaps a black Labrador or exceptionally large feral cat. Occam was open to rational evidence. In our day, for example, spoor, footprints, photographs and fur and blood DNA evidence could alter the Occam default. Likewise, with Occam in mind, the witness might say "I saw a large, black cat-like animal. I think it resembled a black panther (a melanic Asian leopard as in the *Jungle Book*) but I need to gather some evidence before I can say much more." Or, "I saw a long-necked animal in the sea. It looked like pictures I have seen of Plesiosaurs but I can't say that it was a Plesiosaur." Occam might reply that it was something already known: a seal, an eel, a sturgeon, a wave or a product of the witnessing brain, influenced by alcohol, drugs or vivid imagination!

Mystery Animals of the Scottish Borders

MAMMALS
Borders Mammal Miscellany

The Capybara is native to South America but was reported near Eyemouth on two occasions, once in July 2006, sitting in the middle of the busy A1! It was reported again in June 2009.

Wallaby. There are several sightings of wallabies in Scotland. If you Google "Wallaby in the Scottish Borders" you will probably find the photograph of a wallaby filmed by a local retired farmer, in mid-August 2016

River mammals are able to expand easily from outside their introduced range. Two such animals, brought to Scotland for fur farming, escaped and bred in such numbers that they had to be eradicated on a huge scale. The North American Muskrat and the South American Coypu were introduced for fur farming but escaped. Although these animals are thought to be eradicated, it is also possible that breeding populations are still at large.

Mammal

What is it? Capybara
(*Hydrochoerus hydrochaeris*)
The Capybara is the largest rodent in the world, related to guinea pigs and more distantly to chinchillas and coypu.

It is native to South America. It is not threatened or rare in its own habitat. It is hunted by native peoples for its meat, thick skin (from which grease is made) and hide.

What is its natural range?
As its Latin name suggests, it prefers to live near bodies of water, in groups of 10-20 or more. It needs forest cover.

Where has it been seen in the Scottish Borders?
In 2006, a witness reported a capybara sitting in the middle of the A1! A similar animal was reported three years later near Eyemouth.

Where to look for it?
Near Eyemouth?!

What to do?
Report this to local authorities and use social media. Scottish capybara will likely be escapees from zoos or private collections.

Mystery Animals of the Scottish Borders

Comment.
This creature won't be mistaken for anything else - think giant guinea pig. These animals are as big as a large (fat) dog.

Mammal
What is it? Coypu (Myocastor coypus)

What is its natural range?
It is a native of South America and resembles a larger beaver or water vole. It lives near rivers or flowing water, where it burrows and feeds on river vegetation.

Valued for its fur, coypu fur farms were introduced to Britain in 1929 with first escapees recorded in 1932. Eradication took place in the 1930's and 40's but isolated individuals have been sighted since eradication.

Where has it been seen in the Scottish Borders?
It has not been verified in water systems in the Borders, but it has been reported occasionally in water systems adjacent to the Scottish Borders.

Where to look for it?
Along rivers, its habitat is similar to the water vole.

What to do?
Report sightings to police, local ranger and on social media. This creature is similar to water voles, beavers and muskrats so as much tangible evidence as possible would be useful: fur samples, spoor or most likely, good photographic evidence. Note the time and location of the sighting as accurately as possible.

Comment.
Sighting a Coypu in the Borders is very, very unlikely but not impossible. Any walker or amateur naturalist should have foreknowledge of this animal just in case.

Mammal
What is it? Muskrat (*Ondatra zibethicus*)
Related to voles and lemmings, this large semi-aquatic rodent is widespread throughout North American waterways. They are smaller than the similar Coypu. They were introduced into Britain (including Scotland) in the 1930's for fur. They were reared in animal farms, with lots of escapees. An eradication programme took place in 1937 but there have been some rare sightings ever since.

Mystery Animals of the Scottish Borders

What is its natural range?
It is found throughout North American water systems.

Where has it been seen in the Scottish Borders?
There have been unverified reports of animals resembling muskrats but the bulk of fur farming took place in the Central Belt of Scotland, between Glasgow and Edinburgh. This is also where most fur farm escapes took place.

Where to look for it?
Rivers, ponds, lochs. It is semi-aquatic.

What to do?
Report it to local wildlife ranger, go on social media to verify other sightings. Get tangible evidence if possible. Get good photographs as this animal can resemble our own water vole, the coypu and beaver. It might also look like a very big rat (rats are also good swimmers).

Mammal
What is it? Wallaby (genus Macropus, 11 different species)

They are like small kangaroos and are native to Australia and New Guinea.

What is its natural range?
The various species of wallaby have different habits. They are wide-ranging marsupial grazers.

Where has it been seen in the Scottish Borders?
A wallaby was captured on video camera near Belses, west of Ancrum by retired Borders farmer Jim Shanks in mid-August 2016. Google "Wallaby in the Scottish Borders" for the video: it looks like a wallaby to me! Wallabies have been seen all over Scotland and some have established breeding populations in the wild in places such as the Isle of Man, and a small island in Loch Lomond, where a local landowner introduced them in the 1920's. There are several sightings of wallabies in Scotland and the English border counties.

Where to look for it?
They are vegetarian grazers and might share the same habitat as hares and rabbits.

What to do?
Report it to the appropriate authorities. It may be an escapee from a private collection or zoo. Try get a good photo for identification although it might only be confused with a very large mountain hare.

Comment.
Wallabies have been reported from all parts of Scotland, including the city of Edinburgh. They seem to do well in our climate.

Mammal
What is it? Raccoon (*Procyon lotor*)
The raccoon, sometimes spelled racoon, also known as the common raccoon, North American raccoon, northern raccoon and colloquially as coon, is a medium-sized mammal native to North America.

It is an omnivore but spends a lot of time near flowing water, where it feeds on frogs, small fish and insects. It is omnivorous and will scavenge for scraps and edible rubbish. It is notorious for raiding rubbish bins. It stands about 30 cm/one foot in height and weighs up to 10 kilos/23 lb.

What is its natural range?
Its natural range is the North American continent although many animals have been brought to Britain as pets and in menageries and wildlife collections.

Where has it been seen in the Scottish Borders?
In 2014-2015, it was reported at large near Wallsend, Northumberland. It was an escaped pet. This creature could easily cross into Scotland.

Where to look for it?
It is nocturnal, found hunting for aquatic wildlife. It is a good climber and swimmer but requires forest cover. However, like the fox, it will scavenge in public places.

What to do?
Report the animal to local authorities and to the RSPCA. Use social media as it will probably be an escaped pet.

Mystery Animals of the Scottish Borders

Comment.
Like many a North American boy, I tried to keep small raccoons as pets. However, they have a sharp bite, carry parasites and are also escape artists. One of my pets simply reached through the cage mesh and unlatched the hook! They have the delightful habit of washing their food (its German name means "wash bear") before eating, even if the food disintegrates in their dextrous paws. They are delightful animals, canny and perceptive but are truly wild creatures. They should not be in Britain, as pets, or escapees.

Mammal
What is it? Wild Boar: (*Sus scrofa*)

What is its natural range?
The wild boar, always prized for its flesh, was widespread throughout the British Isles. Eventually, it was hunted to extinction. 1260 is given as one date for English extinction and 1541 for Scotland. However, boar were commercially farmed for meat and the first escapees were noted in the 1980's. The great hurricane of 1987 damaged many fences and enclosures, enabling wild boar to escape into the English countryside. Similar escapes happened in Scotland. Now, there are active wild and breeding populations in various parts of Britain. It is a scavenging woodland creature. There are also hybrid varieties which have bred with escaped domestic pigs.

Where has it been seen in the Scottish Borders?
By 2009, wild and breeding populations have been noted in North and Central Dumfries and Galloway. Individuals from that population may be present in other parts of the borders.

Where to look for it?
Isolated forests and woodlands, bordering on rough terrain.

What to do?
Although providing both hunting and a good meat supply, wild boar and associated hybrids can also carry varieties of swine fever and foot and mouth disease, especially relevant to the Borders after the dreaded outbreak in 2001. Report any sightings at once to local authorities. The Science and Advice to Scottish Agriculture (SASA) is especially keen to hear of any reported sightings, particularly in areas where the wild boar is not known. Note the exact time and place. Photographic evidence is especially useful.

Mystery Animals of the Scottish Borders

Mammal
What is it? European Beaver (*Castor fiber*)

What is its natural range?
It was once found throughout the British Isles, becoming extinct in England in Saxon times, Wales in the 12th century and Scotland in the 16th century.

They inhabited rivers and streams with access to woodland.

Where has it been seen in the Scottish Borders?
Norwegian beaver were reintroduced into Knapdale, Argyll in 2009. However, a population of escaped beavers have flourished in the River Tay.

Where to look for it.
Rivers and wetland.

What to do?
Report it to local authorities and any wildlife organisations in your local area.

BIG CATS (The term is often ABC, "Alien Big Cat" meaning an alien or outsider to its natural area. I prefer not to use the term 'alien' because of its extraterrestrial connotations. 'Alien' also presumes to know what the natural range is (or isn't) which presumes more than is presently known.

As a young man, I heard a lot about cougars/pumas/mountain lions on both sides of the American/Canadian Border. One night in the Rocky Mountains I heard one and it made the hairs tingle on the back of my neck. A friend told me that he had once been stalked over several miles by a mountain lion. When he stopped, it stopped. They are canny creatures, very difficult to see in the wild. Moreover, it has been verified that the Eastern North American form of the same animal is alive and well in places such as Florida and Newfoundland. As a North American (born in Canada, lived in the USA) I thought some of my favourite big animals like bears, wolves and pumas would reluctantly be forgotten or filed away as memories. I wasn't long in Scotland though when rumours began to filter in of something familiar. Large cats were present in Scotland, from Caithness to the English Borders. I once saw proof. I went to Inverness and looked at a stuffed animal that had been shot in the Highlands, in Glen Affric. It was none other than an American mountain lion. In this case it was one that had escaped from a menagerie and lived wild for many years on the plentiful game the Highlands provide. However, most cases were less clear. When I worked in rural Aberdeenshire from 1994-1997, the local press was full of accounts of huge 'panthers' roaming the Banff and Buchan countryside. These creatures were seen in daylight and one woman in the village of Aberchirder was menaced by one. Descriptions were consistent. The animal was feline, bigger than a large Labrador dog and nearly always black. I boarded for most of the week in a large farmhouse in the Banff countryside. The owner said he had looked out of the window at the frozen field beyond the house. It was a field of stubble ringed by hardwood forest. It was morning, sometime around New Year's Day in the early 1990's when he saw what he described as 'a black panther with a pheasant in its mouth' It was in no hurry but was walking diagonally across the field towards the ring of trees. He grabbed his shotgun and sprinted to the field. The beast was gone. No tracks or spoor were evident and although he looked often, he never saw the big cat again. Local folk in Scotland are quite open to the possibility of the 'muckle (big) cat.' Like the Beast of Bodmin or similar creatures, the beast was black. The colour itself poses difficulties for the naturalist. There have been no verified cases of black (melanic/melanistic) cougars. Likewise, the famous American black panther has also never been verified. The Black Panther, so popular in fiction and film can only genetically be a colour variation of the leopard or jaguar. The Asian leopards prevalent in India can feature black specimens, rare but possible. The best introduction to this big cat mystery in Britain is found in: Marcus Matthews, <u>Big Cats Loose in Britain,</u> CFZ Press, Devon, 2007. It may surprise many who live here that the Scottish Borders has no shortage of Big Cat reports. Driving west from Galashiels, the Borders changes into a more Highland landscape with high hills, tree cover and a land rich in rabbits and pheasants, birds and small rodents. However, this region is also dependant on sheep. The landscape can be rough and wild so it is no accident that most reports of 'Big Cats' come from places such as Glentress or rural Peeblesshire. Over 60 reports of sightings of mysterious big cats were logged between 2005-2009. This report from the 16 January 2009 *Peeblesshire News* described one such report, which I summarise here:

Near the hotel in Glentress, a jogger discovered a carcass which had been stripped to the bone overnight. He photographed the carcass and also noted that it was the third such sheep death in three weeks. He also photographed some large footprints which proved inconclusive. Mark Fraser of the Scottish Big Cats organisation was quoted: "I certainly would not dismiss a big cat-with the history of sightings in the area, it seems evident that there is a large feline loose, and that has to eat. The print is inconclusive and will need more research and we would need to actually examine the carcass to give a definite yes or no."

The article also notes that several such reports have come in, from Clovenfords to Tweedsmuir. (all rugged and well-protected landscapes)

Earlier in the month, local police investigated attacks on sheep in the Traquair area. These reports and descriptions have remained constant, particularly in the more isolated parts of Peeblesshire. Driving to work in this area, I sometimes never passed another car for miles.

Cardrona Cats
Cardrona, near Peebles, has been the sight of reports of big cats over the years. Although a large white puma-like beast has been reported, black varieties are also described. In June 2002, a black 'panther' was reported. In December of that year a big white cat was reported (the story was picked up by the French press as 'Observation d'un Grand Chat albinos in Ecosse') and two more were seen in February 2003. Local reports stated that experts had been unable to identify the creature. Locally, people began to refer to the 'White Cats of Cardrona' and the name has stuck. Ironically, albinism in cougars is probably less rare than melanism, which is a trait more likely in only jaguars or leopards. Since then, this wild part of the Scottish Borders (hills/forests/rivers/plentiful small mammals and birds) has reported many big cat sightings. The mystery remains.

Galloway Puma
The Galloway Puma is based on a number of sightings of large, black puma-type animals seen most often near Newton Stewart in Galloway. Some Canadian tourists (who would know what a puma looked like) reported their sighting of such an animal. Over the next few years, there were more sightings. Near the Newton Stewart golf course, a dog walker described "a large black cat, bigger than an Alsatian" spring out at her; it was frightened away by her dog.

Kellas Cat
One candidate for the big black cats seen is put forward in a letter to the *Southern Reporter*, 18th July 2002, which also appeared in the *Border Telegraph* of 24 July 2002.

Elaine McCarter and Morag Cameron of Galashiels describe how their own pet cats have gone missing. (five cats in four weeks) In the letter they describe the uniquely Scottish Kellas Cat, based on a cat captured and stuffed on display in the Elgin Museum. Kellas is a town near Elgin. They surmise that the Kellas cat is usually black and follows the breeding patterns of the Scottish Wild Cat. DNA tests show that the Kellas Cat is a hybrid of a feral cat and the Scottish wildcat and is a particularly fierce and elusive hunter. The letter writers appeal to others who may have seen the Kellas Cat in the Borders.

This small dark cat is named for Kellas, Morayshire, where it was first found. It is now believed to be a hybrid between wild and feral cats. Studies of carcasses have shown that the Kellas cat may be a melanistic wild cat, or hybrids closer to wild cats than domestic cats. The animal itself is often between 60-90 cm/2-3 feet and weighing a maximum of 7 kg (over a stone).

Naturalist and researcher the late Di Francis was a recognised authority on the Kellas cat. See: *My Highland Kellas Cats*, Di Francis, Cape, 1993. Most sightings and

Kellas cat. See also illustration on page 67

captures have been in the North East of Scotland thus far. A specimen may be seen at the Elgin Museum and in the University Aberdeen Zoology Museum.

One persistent theory about all Big Cats in Britain is that they are escapees or deliberate releases from zoos, wildlife parks and private collections. This is a sound explanation in some cases. In February 1988 on the Minto Estate near Jedburgh an exotic large cat was killed by a gamekeeper who caught it attacking pheasants. It turned out to be a Leopard Cat that had been bred in Edinburgh Zoo and sold to a private collection in Cumbria, from which it had escaped and survived in the wild for 8 months. A second Leopard Cat was killed in Berwickshire over a year later but its origin was unknown.

Marcus Matthews describes this case in more detail on pages 348-350 of his *Big Cats Loose in Britain.* He quotes gamekeeper William Thomas:

"I saw this creature-it looked like a cross between a leopard and a wildcat-running from a pen with a dead bird in its mouth...the cat had climbed a 10-foot high netting fence to get in. It killed 7 birds-3 cocks and 4 hens..*"*

It was a male cat, weighing half a stone and was thirty inches from head to tail. It had been sighted at other locations in the Borders. The animal was freed in Cumbria as the result of a burglary at the premises. That a young Bengal Leopard Cat could survive eight months in the wild is a good indication that the Borderland is a healthy environment for large felines.

Leopard Cats are native to Asia, including Eastern Russia. They are roughly the size of domestic cats but more slender and long. Their markings differ greatly` from spots to stripes and streaks. They are rarely heavier than 4 kilos. Like the above animals, they are most likely escapees from zoos or private collections.

Lynx (Eurasian Lynx, *Lynx lynx*)
The Scottish Big Cat Trust recorded that a dog walker spotted a Lynx whilst walking dogs in woodland near Coldingham, St. Abbs. The cat had captured a squirrel or rabbit (both abundant in the Borders) It was described as: "some sort of cat it was about the size of a collie dog, medium coat length, it appeared not to have a tail. The dog walker did further research and concluded that it was "some type of Lynx."

The Scottish Big Cat Trust also noted that a Yorkshire man reported a clear sighting of a Lynx in the Scottish Borders "in 1970 or the early 70's. "

Although the Lynx probably became extinct in Scotland in the 13th century, the Eurasian Lynx is still prevalent in wilderness environments throughout Europe, with the largest concentrations in Siberia. It can be 130 cm long and weigh over 30 kilos. It is recognised by its short bobbed tail and black ear tufts. It can vary in colour from reddish brown to silver or black and its markings are also localised. Like many bigger cats, it can make a variety of calls and sounds. They are generally wary and secretive animals but are capable of bringing down relatively large prey such as small deer and larger mammals. The lynx was prevalent throughout Britain in prehistoric times but there are no records of them in historic times. Their fortunes were tied to the destruction of the ancient woodlands of Britain. The most recent date for this feline is around 1250 in Yorkshire. The animal was widely hunted by the Anglo-Saxons into the Middle Ages. There are currently plans to reintroduce the lynx into the Kielder Forest in Northumberland which would eventually result in some animals at least conducting forays into the Scottish Borders so watch this space!

What is it?
Eurasian Lynx (*Lynx lynx*)

What is its natural range?
It is a creature of remote tundra and woodlands. The Eurasian species is native to Siberia, Asia and eastern and northern Europe. It became extinct in the British Isles in the 13th century but many 're-wilding' projects cite the lynx as a possibility. It is

carnivorous, feeding on a wide variety of smaller rodents and mammals, but is capable of killing small deer. It is solitary and requires a huge hunting range. A male will stand up to 75 cm (2.5 feet) at the shoulder. It can be 130 cm/4-5 ft. Long and weigh about 30 kilograms/60-70 lbs.

Where has it been seen in the Scottish Borders?
There have been several reports of animals resembling lynxes seen in the Borders.
In June, 2000 dog walkers reported seeing a lynx which was carrying either a rabbit or squirrel in its mouth. It was described as either a North American or Eurasian lynx which are similar. The sighting was at Coldingham, St. Abbs in Berwickshire. The country was woodland, with lots of undergrowth. In November, 2002, a witness walking a dog, saw an animal "like a lynx" at the Eyemouth golf club at Eyemouth. Three other witnesses also confirmed the sighting.

Where to look for it.
This species is wary and secretive so a walker or naturalist would have to be extremely careful in noise and movement. Any isolated areas with prey species such as rabbits or squirrels would be most likely. However, sightings in the Borders have been in eastern Berwickshire. Rural Dumfries and Galloway would also seem a likely area.

Asian Leopard cat (*Prionailurus bengalensis*)

Mystery Animals of the Scottish Borders

What to do?
Local media might want to be aware of any escaped animals. Social media might be worth a try to discover other sightings or ownership. Local park rangers would want to know as would the police. Lynx are unlikely to harm humans but would fight or defend against a dog. Photographic evidence would be useful. Try find footprints or other evidence. Check out the website for "The Scottish Big Cat Trust." They have a standard form for witnesses to fill in.

Comment.
Seeing a lynx in the Scottish countryside is not so far-fetched. Reintroducing extinct or rare animals is often under discussion. The lynx was once widespread throughout Britain. Scotland might afford many opportunities for re-introduction. The Kielder Forest which forms a large natural boundary with Northumberland has also been discussed for reintroduction of the Eurasian Lynx and other such projects are already underway in Continental Europe.

Wildcats
Although the Scottish Wildcat is rare, it is not likely to be mistaken for any other large cat, even a domestic one. I've only seen one wild cat in the wild, along the verge of the A9 north of Dunkeld. It was about three a.m. on an Autumn evening. I saw the animal pursuing a small mammal along the verge. Its tail was its most distinctive feature. It looked at my slowing car then vanished into a conifer wood. Both its size and tail size were its identifiers. I would not be able to tell if it was a 'pure' wild cat or not, since many of these animals are hybrids (with feral domestic cats)

It has existed for millions of years and predates both humans and domestic cats, with which it breeds, weakening its distinctive bloodline.

Meanwhile, Big Cat sightings have continued:

The Scottish Big Cat Trust did some excellent work in the early years of this new century by cataloguing sightings of unidentified big felines throughout the Borders:

Broughton 2000/2001.
Hawick 2001.

Ancrum 2002
Innerleithen 2002
Melrose 2002/3
St. Boswells 2003.

As we have seen, huge white cats were seen near Cardrona in 2002-2003.

Many letters to the local presses appeared from 2001 to 2003.

Although many of these sightings stem from 2000-2003, a perusal of the *2008 Big Cats in Britain Yearbook* (covering reports from 2007) reports various Borders sightings:

3 February, Berwickshire: A witness reported a completely black leopard-like cat, sighted in late afternoon. It was reported to be nearly one and a half times the size of his own Labrador dog, which was on a lead. It was seen in woodland near a river, where several sheep were grazing.

Dumfries and Galloway produced five reported sightings from January to July:
Three of the sightings were lynx-like and the other two described the large cats as ash grey with markings, and dark tan.

Roxburghshire, Gattonside near Melrose. 28th December. The animal was skulking in roadside bushes. The witness said: "I would say it was a black panther-it had a roundish head as opposed to a sleek one." (p. 216)

On the 4th November, Selkirkshire also produced an eyewitness account: "I would say the cat was 1.5 ft in height and 3-3.5 ft in length. It was seen from 200-150m away for a duration of five minutes."

Finally, the image of a panther seems to re-emerge from this witness's account from Wigtownshire on the 29th of September. (page 219)..."I had a sighting of a cat, it looked like a panther. We were in the car at the time going between Stranraer and Sandhead on the 29th Sept. It was about 12 in the afternoon and we were driving past a wooded area. My eight-year old niece was in the back of the car and my partner was driving. My niece and I were looking for deer in all the wooded areas as we were driving past, when we both saw this. It was very large, black and jumped up on one of the trees in the woods. I hope this is of some interest to you."

The main point here is that it is a remarkable range from the central Borders to the west, in several types of terrain. Again, a dominant image is that of a black panther.

Many things are remarkable about Big Cat sightings in the Scottish Borders. Some witnesses are quite clear that they are seeing a lynx. Perhaps the colour range in the

Borders is also remarkable, ranging from the 'white cats' of Cardrona to the tawny brown specimen at Ancrum with many describing a black panther-like animal elsewhere. Moreover, although most sightings have been in the wilder and more remote parts of the western borders (chiefly rural Peeblesshire and rural Dumfries and Galloway) many sightings have been in the more populous eastern areas like Melrose and Galashiels. The origin of these big cats is the biggest mystery which needs solved with photographs or films of the beast itself, its footprints and habits. DNA is desperately needed from blood or fur. Although these big cats are certainly secretive and elusive, that secrecy would end with a carcass. The numerous sightings along busy roads at night (where road kill is abundant) mean that it might only be a matter of time before one of these mystery animals itself becomes a road victim. The biggest mystery of all is why are they so often black, when melanism is rare among big cats, found only rarely in African and Asian Leopards and South and Central American jaguars. Although the smaller Kellas Cat is often tawny or black, this native creature is much smaller than the animals being reported and would not be able to take down an adult sheep or deer. To date, no melanic/melanistic North American puma (cougar/mountain lion) has ever been confirmed.

The Romans had a fondness for African black leopards in the gladiatorial ring. Did they bring these animals to Scotland with them for gladiatorial combat or to subdue native tribes then release them over time? This mystery can only be solved with hard forensic evidence and not wild speculation. Huge black cats have been seen by reliable witnesses all over the British Isles and it seems the Borders have not been left out.

IN THE FIELD
(Animals: highly unlikely, highly improbable, BUT not impossible)

(This section is gratefully based on an idea from *"Mystery Animals of the British Isles: The Western Isles* by Glen Vaudrey CFZ Press, 2009) I have included unusual animals that have been seen in the Borders or could possibly be seen based on past information. This list is only a sample and far from inclusive.

What is it? Leopard Cat
(*Prionailurus bengalensis*)
Length: Adult. 70 cm/2.5 feet
Weight: Adult. 4 kg/9 lb.

What is its natural range? It is a native of Asia (south/southeast/east) but is kept in British zoos and private collections. It has a wide distribution and is not considered endangered but its native habitat is being destroyed, particularly by logging and general deforestation. Human population pressure is felt keenly in its natural range.
Where has it been seen in the Scottish Borders? In February 1988 a gamekeeper killed a leopard cat which was preying on estate pheasants. A second leopard cat was also killed in Berwickshire a year later.

Mystery Animals of the Scottish Borders

Where did it come from? The first specimen was born at Edinburgh Zoo and then sold to a private collector in Cumbria. It escaped and lived in the wild for eight months before it was killed.

Comment: What this does prove is that 'big cats' can thrive in the Borders and pheasants aside, there seems to be enough prey for them to flourish. The above creature had survived for eight months and was far from its Cumbrian home.

What to do? Note the details of the sightings. Location, time of day, what it was doing, perhaps where it was headed. Get a photo if possible and note any other tangible evidence like fur or spoor. Report it to any local estates or landowners, to local rangers and to the police. Go on social media to report the sighting because that might suggest where and how it escaped. The leopard cat is not considered dangerous to humans but caution approaching it would be advised, especially in tight quarters.

What is it? Cougar/Puma/Mountain Lion. (*Puma concolor*)

What is its natural range?
Its natural range is the Americas, from northern Canada, to the Southern Andes in South America. Black versions are yet unknown, and albinism is extremely rare. This big cat is only slightly smaller than the jaguar, a near relative. Its prey is varied but includes mammals both small and large and it will also take livestock. It preys on small deer. When hungry, it will also eat insects and small rodents. It is reclusive and prefers to hunt at dusk or at night. It is highly territorial, solitary and requires a huge range.

Where has it been seen in the Scottish Borders?
Cougar-like animals have been reported from nearly every part of the Borders but spanning many years. Sightings have come from more remote areas such as Glentress, Cardrona and other parts of Peeblesshire. There have also been many reports from Galloway.

Where to look for it?
Glentress, Cardrona and various parts of Galloway.

What to do?
Note the details of the sightings. Location, time of day, what it was doing, perhaps where it was headed. Get a photo if possible and note any other tangible evidence like fur or spoor. Report it to any local estates or landowners, to local rangers and to the police. Go on social media to

Mystery Animals of the Scottish Borders

report the sighting because that might suggest where and how it escaped.

Comment:
The cougar is wary of humans but attacks on people have been recorded. In its native habitats in North America, it has also been known to stalk humans.

What is it? Kellas Cat (*Felis silvestris+Felis catus*)
This small dark cat is named for Kellas, Morayshire, where it was first found. It is now believed to be a hybrid between wild and feral cats. Studies of carcasses have shown that the Kellas cat may be a melanistic wild cat, or hybrids closer to wild cats than domestic cats. The animal itself is often between 60-90 cm/2-3 feet and weighing a maximum of 7 kg.

The late Di Francis was a recognised authority on the Kellas cat. Most sightings and captures have been in the North East of Scotland thus far. A specimen may be seen the Elgin Museum and in the University Aberdeen Zoology Museum.

What is its natural range?
Like the wildcat, it prefers a lot of tree cover. It is probably similar in habits to the Scottish Wildcat and feral cats. Its natural prey would be small mammals (including domestic kittens) such as voles, mice, rats, etc. It could also prey on rabbits, birds and

Mystery Animals of the Scottish Borders

squirrels, large insects and worms.

Where has it been seen in the Scottish Borders?
In 2002 there was a spate of domestic cats gone missing from the Parsonage Road area of Galashiels and one of the owners put the Kellas Cat forward as a suspect, making a strong case. Although many black panther-like animals have been reported in all parts of the Borders (from the North Sea to the Solway Firth) it seems unlikely that the much smaller Kellas Cat is responsible for these 'big cat' sightings.

Where to look for it?
Forest and remote hill habitats, at night or dusk.

What to do?
Report the sighting to relevant authorities: police, local park ranger, RSPCA. Use social media to confirm similar sightings. Try to gather footprints or organic evidence. Get some good clear photos! The discovery of a Kellas Cat or cats in the Borders would be useful to researchers and scientists.

Comment.
The Kellas Cat is wary and of no threat to humans. However, it could prey on family pets. It is not known in the Scottish Borders but the local habitat is ideal for such a predator.

What is it? Scottish Wild Cat (*Felis silvestris silvestris.* Some argue *Felis silvestris grampia* depending on whether the Scottish Wildcat is a species or sub-species .

What is its natural range?
It was once found throughout Britain and in all parts of Scotland but historical persecution has restricted it to remote parts of the Highlands and Grampians.

Where has it been seen in the Scottish Borders?
It has been reported in all parts of the Borders but no positive confirmation has been made.

Where to look for it?
It hunts by night and would be found where there is much tree cover and where there is an abundance of prey (small mammals) Remote parts of Galloway and

Mystery Animals of the Scottish Borders

Peeblesshire would be the most likely areas to support this secretive animal.

What to do?
Any Borders sightings would be valuable for professionals. Notify any wildlife agencies. Get photographic evidence and look for organic evidence. Uncontaminated DNA would be ideal. Note the time and any other details of the sighting.

Comment.
A confirmed sighting of the Scottish Wildcat in the Scottish Borders would be invaluable. There have been many reports of lynx-like animals in the Borders but the lynx has been extinct for centuries. Although Wildcats are rare, they are not extinct but must be considered to be more probable sightings than that of the Eurasian Lynx.

Big Cats

What is it? Black Panther
Many unidentified big cat sightings throughout the British Isles (over many decades) refer to a black animal, the size of a cougar. This poses a dilemma because black panthers can scientifically only be melanistic leopards (Panthera pardus) or jaguars (Panthera onca) No black cougar/puma/mountain lion has ever been photographed, shot or trapped. Frequent sightings of black panthers in America are thought to be black jaguars whilst in India and parts of Asia, they are black leopards.

What is its natural range?
Leopards are found in Asia and Africa whilst jaguars are found in Central and South America. It is a carnivore, a good climber and is generally solitary. It requires a big range, feeding on smaller mammals and birds. It prefers natural cover.

Where has it been seen in the Scottish Borders?
Everywhere, particularly in the western counties. Here are some sightings: Preston, Cardrona, Portmore Loch, Melrose and many other areas.

Where to look for it?
Wilder, more remote parts of the Borders such as Glentress, Cardrona, and Galloway.

What to do?
Do the same as for cougar/puma/mountain lion: note the details of the sightings. Location, time of day, what it was doing, perhaps where it was headed. Get a photo if possible and note any other tangible evidence like fur or spoor. Report it to any local estates or landowners, to local rangers and to the police. Go on social media to report the sighting because that might suggest where and how it escaped.

Mystery Animals of the Scottish Borders

Comment.
Panthers are stealth hunters and could be dangerous to humans, livestock, dogs and other pets. One theory is that Romans kept leopards for gladiatorial combat or as status symbols. Black leopards were particularly prized. If they brought these animals to Britain (the Romans were once widespread in the Scottish Borders) then when the Romans left they would have released their pet black leopards, thus releasing a population of big cats with a black gene pool.

INTERLUDE: ANIMALS IN THE BORDER BALLADS

Lowland Scots/Lallans/Border Scots is a Germanic language that very early diverged from English. It is related to English the same way Spanish is related to Portuguese or Norwegian is to Swedish. It was the preferred tongue in the Borders, especially amongst rural folk. Scots is a first cousin of English, *not* a dialect of English. Borderers speak it and know it with pride. Old Borders place names reflect this: *Bemersyde* refers to the Bittern, no longer a common bird in the Borders. *Cranshaws*=Crane, *Gled's Nest*=the Red Kite and intriguingly, *Wolf Cleugh* (cleuch/cleugh=steep ravine) when wolves roamed the local woodlands centuries ago. *Pyatt Field*. The 'pyatt' locally is a magpie. There are many more examples on the map. Scottish Gaelic was especially spoken in the wilder parts of Galloway, which gives the area its name and some its natural place names have Gaelic origins.

The Border Ballads are the great art form to emerge from both sides of the Border. These ballads chronicle events from 1500-1700, owing to both the politics and culture of the Reivers. Reivers were cattle raiders (on horseback) who dealt in blackmail, kidnapping and extortion. In a nutshell, a Borderer crosses the border and plunders his rivals. If you are caught, your comrades may free you, like Kinmont Willie from Carlisle Jail. There are also plenty of supernatural happenings. Witches and sorcerers abound. The men and women respect the finest horses and cattle, and appreciate good hunting as well. I thought it might be fun and instructive to see what animals they might have held in high esteem. What follows is not a checklist of Borders fauna. Probably, to the people of the time, thieving and

Mystery Animals of the Scottish Borders

fighting were more interesting than the wildlife, which wisely stayed hidden when the 'Mosstroopers' rode out! Although all Reivers were Scottish or English borderers, not all borderers were Reivers since most of the rural population were not wealthy enough to own fast horses, armour or expensive weaponry.

(Reference: **Border Ballads.** William Beattie, Penguin, 1952)
Some wild animals mentioned in the Scottish Border Ballads:

- Chevy Chase. "fallow deere"
- Jamie Telfer of the Fair Dodhead. "mouse" hawks"
- Kinmont Willie. "Corbie's" nest (in the Borders, a crow is a *corbie*)
- Tam Lin. "Esk/adder/bear/lion"
- The Twa Corbies (the two crows) "hawk"
- John Hewit, blacksmith and bird watcher from Heiton, 1956: "they change their voices every month and as the season comes round, you can tell aforehand wi' the cry o' the craws. You can notice the change." from: *Animals and People in Scotland.* Pullar, Polly. Low, Mary. Birlinn, Edinburgh, 2012.
- 'Jock o' Hazeldean' "Nor mettled hound nor managed hawk..." The managed hawk is probably the merlin from the Latin merula/ blackbird. This small falcon eats other birds and was popular with falconers.
- Proud Lady Margaret. "And what is the bird, the bonnie bonnie bird, sings in the evening gale?" That bird is the "thistlecock."
- The Broom of Cowdenknowes. "tod" (fox)
- The Gay Goss-Hawk. "Goshawk"
- Kempion. A mysterious "fiery beast" (a dragon?) Also mentioned the warwolf, the mermaid
- Lord Thomas and Fair Annie. "Seven young rats/seven young hares".
- Edward, Edward. "O, I hae killed my hauke (hawk) sae guid (so good)"

Note that one ballad features the Goss-Hawk/Goss Hawk (*Accipiter gentilis*) an efficient and powerful hunter and killer, and one of the most difficult birds for falconers to train. It's possible that the population of Goshawks in the Scottish Borders (it is relatively rare in the British Isles) was a result of escapees from the falconer's art, since it was previously hunted to extinction. Its hunting prowess and fearless attack make it a good symbol for that lawless time in the Borderlands. That other eponymous bird in the Borders (Corbie/crow) is the favourite prey of the lethal

Goshawk. Any bird that can take on a Borders Corbie is a bird to be reckoned with and well worthy of a ballad!

Dandie Dinmont Terrier (not from a Border Ballad but from Sir Walter Scott's novel, *Guy Mannering*

Although many mysteries pose unanswered questions, here is the answer to what sounds like a classic pub quiz question. What is possibly the only recognised breed of dog that takes its name from a fictional character? Answer: the Dandie Dinmont Terrier, whose name comes from a fictional character in Sir Walter Scott's novel *Guy Mannering* (1814). In the novel, Dandie Dinmont is a rough-hewn farmer in the Scottish Borders who owns several terriers, all called *Mustard* or *Pepper!* I'll paraphrase some further information from *The Observer's Book of Dogs* by Clifford L.B.Hubbard, London, Frederick Warne and Co. Ltd, 1945:

The breed is also called 'Charlie's Hope Terrier' or 'Mustard and Pepper Terrier'. It originated in the Borders (English or Scottish) in the 18th century. The name was adopted in 1820. The Dandie Dinmont Terrier Club was established n 1875. The dog stands about a foot tall and weighs 18-24 pounds. Its colour is "pepper or mustard, tawny red or dark grizzle. Coat soft yet thick." It has strong jaws. It also sports a very distinctive top knot of hair. On both sides of the Border the breed was used in early centuries to hunt badgers and otters, and more rarely, foxes.

It was registered with the American Kennel Club in 1888 and the world-wide United Kennel Club in 1918. It is now considered one of the rarest breeds (most Borderers have never seen one and many have not yet heard of it) and is on a list of 'Vulnerable Native Breeds'. Those not knowing the dog might confuse it with the more common Skye Terrier.

Most recently, the Haining in Selkirk (The Southern Reporter, 8 June, 2017, Kevin Janiak) brought the Dinmont Dandie legend full circle with the unveiling of a sculpture by Sandy Stoddart. Stoddart's sculpture was of Old Ginger, said to be the 'father of the Dandie Dinmont' breed. 125 Dandies turned up for an event to honour the iconic Borders breed. The article emphasised the rarity of this terrier, with only 316 born in the world in 2016.

The Dandie Dinmont has a lot of the classic Terrier qualities: hardy, friendly, loyal, protective, brave and a good companion generally. They can be expected to live 12 years or more. Likewise, owners are very loyal to the breed.

ZOOFORMS

This term was coined by Jonathan Downes, the indefatigable author, lecturer and explorer who is a renowned British cryptozoologist. My own take on Zooforms is that there might be a supernatural or unnatural aspect to these creatures but that the term encompasses all the animals that we *know* but have never actually seen: kelpies, unicorns and mermaids as well as more possible and mysterious creatures like big cats, bigfoot, the yeti and so on. Anyone afoot in the Borders can easily understand the appeal for artists and poets: mountains, rivers, unspoiled terrain. It's no wonder that these imaginary creatures thrive here.

Again, I go to Alan Temperley for two cracking examples (from *Tales of Galloway*)

The Mermaid of Barnhourie

A young mermaid of the Solway Firth falls in love with a young fisherman from Dalbeattie. However, the fisherman is happily married with a wife and children of his own. After his boat is destroyed in a storm, the mermaid saves his life. They fall in love and he fathers her mermaid child. He returns to his own family but after a second shipwreck, the mermaid is unable to save the young fisherman who drowns. However, a few secret potions enable him to live on as a merman, with the young mermaid (mermaids don't age) and their son, who has his mother's lovely tail and the

man's wavy hair and good nature. His land wife and children are taken good care of and his widow re-marries happily. The mermaid and her husband then save many fishermen from watery graves and continue to do so to this day!

The Kelpie or Water Horse is a real shape-shifter, most often associated with the Highlands. It is said to kidnap beautiful girls and can appear in many forms: goats, frogs, horses. As a horse, it will take its rider into deep water where the rider will be drowned and/or eaten. In the West of Scotland they can appear as otters, whales or massive fish. Again, from Alan Temperley, we are offered a west Borders version of the ancient story. A traveller must cross a stream in spate but when he dislodges a stone he disturbs a kelpie. She is a beautiful woman naked but her long hair, entwined with newts and eels covers her modesty. She warms him that he will die unless he crosses the stream before daylight. However, his limbs fail him throughout the night and he is unable to cross. Just before dawn, his strength returns just in the nick of time and he crosses to the opposite bank. The image of the kelpie haunts him for the rest of his life and he is also aware that he has had a narrow escape.

The Border between Scotland and England is marked by natural features. The Tweed is shared by both ancient nations. It is the place to list a few mythical creatures from across the Border that could easily slip into Scotland by river or forest. Scotland shares the longest land border with Northumberland, bordering on the wild and remote Kielder Forest Park as well as the Northumberland National Park. The Tweed forms only a small part of that border and the border is a mystery to those not living either side of it. For example, Berwick-upon-Tweed is in England but has been Scottish on 12 previous occasions! It is also the unofficial capital and chief market for

Berwickshire, which is entirely in Scotland. To get to Berwick, I must actually drive north to get to England! I can drive nine miles to Coldstream in Scotland, cross the bridge into England and drive home most of the way in England. Crossing the bridge both ways, one sign simply says "England", the other "Welcome to Scotland". However, although on the English side you aren't told you are *Welcome to England*, you are told that you indeed *Welcome to Northumberland.* In most places, when I cross the Tweed from Scotland, I am still in Scotland, but there is a small border where crossing the Tweed takes me from Scotland to England. Like most borders composed of natural features like forests, mountains and rivers, wildlife crosses freely. A naturalist must therefore be aware of the real and imagined wildlife on both sides of the border. Fortunately, Michael J. Hallowell has written *Mystery Animals of the British Isles, Northumberland and Tyneside*, published by CFZ in 2008. From the vast and varied Northumbrian landscape the author has given us giant rabbits, fierce black dogs, panthers, vampire and witch rabbits, mermaids, monster lobsters and worms/orms/dragons similar to the Linton Worm!

Just as wolves, grizzly bears, caribou and geese travel freely from Canada into the States by cover of mountain, forest and darkness so might some of those wonderful Northumbrian beasts cross to and fro. I'm hoping that my own modest book repays some of that cross-border debt!

Likewise, from Neil Arnold's *Monster: THE A-Z OF ZOOFORM PHENOMENA* I note a hat-trick of remarkable creatures from neighbouring Cumbria.

Cumbrian Cockatrice.
A cockatrice is a hybrid creature with a long pedigree. It is said to have the head and legs of a cockerel and the body and tail of a dragon. It can kill with just a stare. The cockatrice can only die if it stares at its own reflection. It was sighted in 1733 over the skies of Cumbria. My own ancestors (Goodfellows) were farm labourers in Cumbria at the time and I wonder if they saw it! It would surely also fly into Scotland from time to time.

Cumbria Bigfoot. A tall, hairy biped (not human) was seen drinking from a pond at Nursery Woods, Bekermet in January 1998. The same or similar creature was also seen on 1 October 2005.

Windermere White Horse. Rarely seen, this creature is considered an omen.
Not to be outdone by our Cumbrian neighbours, we hae a few beasties of oor ain!

Water Bulls at Cauldshiels and Alemoor Lochs.
The dangerous shape-shifting *kelpie or water horse* is thought to be mainly a Highland creature. However, the Scottish Borders can claim a similar aquatic shape-shifter (also found in the Highlands). This is the mysterious *tarbh uisge*, meaning 'water bull' in Gaelic. Before we learn where to find them we need to learn more about them. Water bulls can breed with 'normal' land cattle, often producing beasts of

superior quality. Water bulls are best got rid of at birth but can't be drowned. Like the vampire, it can be killed with silver or silver bullets. It can assume many forms (making classification difficult indeed!) It can live on land but prefers water. They don't have ears hence any breeding with a normal cow produces calves with small ears. This creature is also capable of acts of kindness, having once saved a maiden from a fearsome water horse (kelpie) suggesting that these two shape-shifters are natural enemies. Water bulls are thus not as dangerous to humans as kelpies.

Although many larger Highland lochs boast *both* kelpies and water bulls (there are many water bull stories from St. Kilda) there are two very specific lochs in the Borders associated with the water bull.

The first is the lovely Cauldshiels Loch near Melrose. The beast was known to locals for centuries and Sir Walter Scott wrote of it. Many writers and poets in the Borders had heard of the beast in the loch.

The creature living in the small Alemoor Loch west of Hawick was/is either a classic kelpie or else a water bull like that of Cauldshiels Loch. Alemoor Loch, now popular with anglers, is more likely to be the home of pike and perch rather than a mythical creature, However the loch had a very different depth and shape before it was turned into a reservoir. It was said to be nearly 200 feet deep in places, ideal for the water bull of legend. As far as I know, there haven't been reports of water bulls in either loch since Victorian times. Both lochs are worth a visit. Let your imagination do the rest. If you aren't imaginative then take delight in the very real creatures you may see in both places. Dragonflies have a wonder and magic all their own.

UNICORNS (PARTICULARLY THE 'SCOTTISH' ONE)
The lion and the unicorn were fighting for the crown,
the lion beat the unicorn all around the town.

18th Century child rhyme
If Zooforms are mythical creatures that may have once seemed real, then the unicorn is right top of the list, even in Scotland. It is a white, horse-like (sometimes goat-like) beast with a long single horn, often featured in paintings. Educated people believed in this creature until the late nineteenth century at least and wiser country people (including those in the Scottish Borders) may have held that belief even longer. Here are some unicorn observations:

The lion is a symbol of England, the Unicorn of Scotland. It also appears on coats of arms in Canada. In the Scottish arms, the unicorn appears on the left, on the English coat of arms, it appears on the right.

The unicorn is a symbol of purity and grace.

It also symbolized a creature that would rather die than face capture

Mystery Animals of the Scottish Borders

It can only safely be captured by a female virgin. It will put its head on her lap then fall asleep!

Unicorns were honoured in Scotland on the 9th April, 2017 on 'National Unicorn Day.' A large willow unicorn sculpture was created by artist Woody Fox for display in the Scottish Borders.

It goes back to the 12th century as a Scottish symbol, partly because it was feared and respected by that other (mythical) English creature, the English lion.
Images of the unicorn have been displayed for thousands of years, from some of the earliest civilisations.

Its powdered horn was said to purify water and heal many ailments. This is probably the basis of the cruel treatment of the rhinoceros whose horn is made of the same stuff as our toenails, which aren't coveted by any poachers I know of!

In medieval symbolism, it is often linked to the Virgin Mary. Later, it symbolized chastity and fidelity

Many animals suggest some of the features of unicorns: the rhinoceros, the narwhal (a cetacean with a single horn) a type of antelope called the oryx, native to Arabia, and the South African eland or any variations of more common animals like roe deer with horn deformities, including a fused pair of horns or antlers.

Unicorns. Both James IV and James V issued coins called unicorns. In the Treasury Accounts of 1488, we learn that the Cross Kirk in Peebles was awarded some such coins:

"To the King, to offir in the Cors Kyrk of Pablis, one unicorn..."

Shellycoat

Faeries, brownies, goblins and such-like are fairly universal creatures, not linked specifically to Scotland or the Scottish Borders. However, some are and the Shellycoat is one. This goblin-like creature is mentioned in many Scottish poems. Walter Scott wrote about him. This is my account:

Come with me on a summer's walk along the Ettrick. It is early evening. Mayflies hatch on the surface. Trout rise for them. More mayflies come, more trout rise. The river is always moving. Damselflies and dragonflies skim the surface. Then, on a huge rock on the far bank I see him. A Shellycoat...who is well-known in the Borders and along the Ettrick. He is about the size of an otter, human-like but not entirely. His most distinctive feature is a coat of shells, which clatter as he sneers and dives into the swirling water, among the trout and mayflies. After a few whiskies in the *** Bar in Selkirk, I used to see Shellycoats regularly. Here's what locals told me about them. They shout "help, help" or "lost, lost..." taking great mischief in bringing out search parties who search in vain for the seemingly lost or drowning person.

Some say the Shellycoat is monstrous, much bigger out of water than he appears (*he is always a he*)

He is fast, likes blood.

His coat rattles but can be removed. When coatless, the Shellycoat is harmless.

There are rhymes that can be learned to ward off the Shellycoat's mischief but these same rhymes might also invoke the Shellycoat rather than repel him. Shellycoats will drown you or at least give you a good hiding. Shellycoats can lift a man up and toss him about like a rugby ball

Many in the Borders believed that the Shellycoat was merely a human under a spell, which turns it into this river goblin. Like many a Scottish ghost, it can be foiled by crossing running water, which Shellycoat does not like to do.

Part of the Shellycoat's character is that it won't harm a pregnant woman.

Like the rattlesnake, you'll hear a Shellycoat before you see it because its rattling coat is very noisy. Once, on a very deserted Highland river bank, when my back was turned I could hear a very distinctive rattling behind me. Shellycoat? No, stonechat.

Bogles

Every Scot knows that a Tatty Bogle is a scarecrow in a potato field. Perhaps not many know that the Bogle is one of the many mythical creatures shared over the border. The selfsame creature is described in Northumbria. Bogles have long been invoked by Scottish mothers to change unruly behaviour in children. Knowledge of scarecrows is more specific than knowledge of bogles. We know they are human in form, throw things but are more mischievous than malevolent. Like dogs and cats, they are domesticated and enjoy confusing us. More field work is needed on bogles. How big do they get, what do they eat, what colour(s) are they, and do they speak Scots or Northumbrian (or both)?

Red Cap/Redcap

These creatures are found on both sides of the border. Unlike Dunnies, Redcaps are murderous and get their name from their victim's blood, used to stain their caps red. They need to keep murdering because legend has it that if the blood on their caps dries, they too will die. Telltale signs:

Redcaps are very fast
Redcaps have heavy, iron-shod boots
They carry very heavy iron pikes
They look like old men
They have red eyes and talons
They have sharp teeth
Redcaps are popular in films, video games and fiction

There is debate about the species: goblin, dwarf, fairy, elf?

No man, not even Usain Bolt, can ever outrun a Redcap!

They seem to have no preference about their country of abode but rumour seems to have it that Redcaps are more Scottish than English and are happiest in the Scottish Borders. I have no advice on what to do when you encounter one. Maybe they can't climb or swim, but don't try outrun them. Remember also, they die if the blood on their caps dries out. Perhaps a battery-powered hair dryer might be the answer (or fire!)

Dunnie
(a species of Brownie)

In my native Canada, animals crossed the Canadian/USA Borders more freely than humans did. Generally, wild creatures seek warmth and more or better food. Wolves, eagles, caribou and geese were thus often stateless creatures. Likewise, whatever our political future, any animal in England or Scotland can cross rivers, forests and mountains whenever it chooses. If this is true of real creatures, it is probably doubly true of imaginary ones. The Dunnie would be one of these. The Dunnie is a form of Brownie, a small human-like creature. Brownies are wrinkled and they have curly hair. They dress in brown, hence their name. Brownies will live in your house and even help with some of the household chores, which they prefer to do at night when the house is asleep. Leave honey or good porridge out overnight and maybe your kitchen will be swept and dusted. Be warned, Brownies take the huff easily and your gifts of food had better be up to standard--- or else.

Brownies prefer quiet, unused parts of the house and might be heard in the attic or cellar.

Tethered mainly to Northumberland, this shape-shifter usually prefers the English side of the Cheviots. It is a classic type of Brownie who turns into an obliging horse that will likely leave its rider in mud or water. If it becomes a plough horse it vanishes from its ploughman, even when in harness. Many consider the Dunnie the ghost of an English Reiver (Border raider) who travels to protect his various hidden treasure troves. The most famous version of the Dunnie hails from the Parish of Chatton in Northumberland.

So Scottish Borderers beware, a too obliging and obedient horse might leave you stranded in the middle of the country at night, on his way back to Doddington, Tweedmouth, Spittal or Holy Island:

"Of aa the toon w'er I saw, Holy Island for need, Holy Island for need..." sings the Dunnie.

This creature is more of a nuisance than a danger and is probably otter-like in his elusive and canny behaviour. He is not averse to causing mischief both sides of the border. Perhaps he (or she) is a politician!

Not to be outdone, Arnold's book also lists a hat-trick from Scotland's other border with Northumberland.

- **Hexham Werewolf.** Seen in summer 1975, in Hexham, after the discovery of two strange stone heads found in the garden of a local family.
- **Northumberland Yeti.** Parallels with the Cumbria Bigfoot. This hulking hairy hominid was seen near Bolam Lake by several witnesses, in 2002 and 2003.
- **The Toad Man of Chillingham Castle.** When excavation was taking place, a giant toad sprang from the crumbling walls, transformed into a man, then vanished!
- We now switch the action back to the Scottish Borders:

- **The Red-Breasted Swans of Closeburn Castle.** Closeburn Castle is north of Dumfries and once hosted a pair of swans. They were omens of good luck until a teenage boy killed one of the birds with an arrow. Since then, their appearance is an omen that someone in the Kirkpatrick Family of Closeburn Castle will die.

- **The Yellow Ape of Drumlanrig.** This apparition of a yellow ape has been seen for over three hundred years in the grounds of Drumlanrig Castle, near Thornhill in Dumfries and Galloway

Meanwhile, here are three recent updates proving that Borders life, both real and imaginary, is always on the move.

Rare Greenland sharks off the Borders coast?
7 October 2013.
The Greenland Shark (*Somniosus microcephalus*) is a wondrous beast. It is a flesh-eating shark that can live at depths of over 2000m and may live to be 200 years old or older. It grows up to 26 feet long. In its environment it feeds on fish, seabirds, seals and even polar bear carcasses. It is caught around Norway, Iceland, Canada and Greenland. Thus it was a surprise when in Ocober 2013, a nine foot specimen was found on a beach outside Alnwick, Northumberland by a farmer who was out surfing

that day. How did it get there? Was it thrown overboard as bycatch by a returning trawler or has global warming affected the habits of this rather scary and mysterious denizen of the deep?

The Southern Reporter of Thursday, 21 September 2017 reported a five foot long iguana found in a Borders layby! It was found by a school janitor in a layby near Blyth Bridge. Nicknamed Iggy, the Green Iguana was safely rescued and was recuperating until his owner came forth or until the reptile could be housed happily!

Elsewhere in this book, credit has been given to the many women and men who work selflessly to monitor and record every animal species in the Borders. *The Scotsman* of 15 February 2018 announced: 'Rare butterfly in Scotland for the first time in 130 years..." the White-Letter Hairstreak produced a clutch of eggs on Wych Elm (Dutch Elm disease has played a part in its rarity because this species feeds on varieties of elm) The clutch was found at Lennel, near Coldstream in the Scottish Borders. Borders butterfuly recorder Iain Cowe had seen a living specimen the previous summer, north of where the eggs were discovered. This discovery was a team effort, which included Ken Haydock and Jill Mills, both volunteers. The meticulous observation and knowledge of those who made this discovery offer proof of the value of dedicated amateurs and volunteers.

Bibliography

- Arnold, Neil. MONSTER! THE A-Z OF ZOOFORM PHENOMENA. CFZ Press, Bideford, North Devon: 2007.
- Arnold, Neil. Mystery Animals of the British Isles: Kent. CFZ Press, 2008.
- Borders Bird Report No 27. Scottish Ornithologist's Club. Borders Branch.2011. Edited by Ray Murray.
- Coleman, Loren and Clark, Jerome. Cryptozoology A to Z. Fireside, New York City, 1999.
- Damselflies, Dragonflies and Butterflies. Margaret Carlaw and Derek Ogston. Ballieknowe Publishing, Stichill, Kelso 2010. (Wildlife in the Scottish Borders Parish of Stichill)
- Dupre, Ben. 50 philosophy ideas you really need to know. London: Quercus, 2007
- The Fish Species of the Tweed and Eye Catchments. Tweed Foundation. various updates.
- Mystery Animals of the British Isles: Northumberland and Tyneside, 2008. Michael J. Hallowell.
- Fraser,Mark. Editor. Big Cats in Britain: Yearbook 2008. CFZ Press, 2008.
- Gerlach, Justin. Extinct animals of the British Isles. 2014.
- Heuvelmans, Bernard. In the Wake of the Sea Serpents. Rupert Hart-Davis, London 1968.
- Heuvelmans, Bernard. On The Track of Unknown Animals. Rupert Hart-Davis, London 1958.
- Mautner, Thomas. The Penguin Dictionary of Philosophy. London, Penguin. 2005.
- Moths. Margaret Carlaw and Derek Ogston. Ballieknowe Publishing, Stichill, Kelso. 2010. (Wildlife in the Scottish Borders Parish of Stichill).
- Peeblesshire News. 16 January, 2009.
- Robinson, Dave and Groves, Judy: Introducing Philosophy: A Graphic Guide to the History of Thinking. Icon Books, 2007.
- Scottish Federation for Coarse Angling: Scottish Record Coarse Fish List.

- (updated as required)
- Steel, Judy (editor) A Shepherd's Delight: A James Hogg Anthology. Edinburgh, Canongate, 1985.
- Temperley, Alan. Tales of Galloway. Edinburgh, Mainstream, 2001.
- The Southern Reporter. Weekly newspaper.
- Vaudrey, Glen. Mystery Animals of the British Isles: The Western Isles. CFZ Press, Bideford, North Devon, 2009.
- Yarrell, William. A History of British Fishes. 1841.
- 2016 Yearbook. Edited and Compiled by Jonathan and Corinna Downes. CFZ Press, Bideford, Devon.2016.

Appendix 1

Addresses and Contacts

Scottish Natural Heritage Offices
Dumfries and Galloway Area Office
Carmont House
The Crichton
Bankend Road
Dumfries DG1 4ZF
(A free poster of the Natterjack Toad is available from this office)

*

Dumfries
Greystone Park
55/57 Moffat Road
Dumfries DG1 1NP

*

Galashiels
Anderson's Chambers
Market Street
Galashiels TD1 3AF

*

Newton Stewart
Holmpark Industrial Estate

New Galloway Road
Newton Stewart
Wigtownshire DG86BF

*

The Scottish Wildlife Trust
Cramond House
Cramond Glebe Road
Edinburgh EH4 6NS

Scottish Wildlife Trust
Reserves and/or Visitor Centres in the Scottish Borders area (special wildlife in parentheses)

(https://scottishwildlifetrust.org.uk)

- Bemersyde Moss (wintering wildfowl/otters)
- Blackcraig Wood (birds/butterflies)
- Carsegowan Moss (merlins/owls.harriers)
- Carstramon Wood (red squirrels/migrating birds)
- Drummains Reedbed (wildfowl/sedge warblers/reed buntings)
- Duns Castle (wildfowl)
- Feoch Meadows (butterflies)
- Fountainbleau Ladypark (wildfowl/butterflies)
- Gordon Moss (birds/butterflies)
- Hare and Dunhog Mosses (wildfowl/dragonflies/damselflies)
- Hoselaw Loch/Din Moss (wildfowl)
- Knowetop Lochs (butterflies/dragonflies)
- Southwick Coast (birdlife)
- Stenhouse Wood (breeding birds/red squirrels)
- Whitlaw Wood (varieties of birdlife)
- Yetholm Loch (breeding birds/otters)
- The Wildfowl and Wetlands Trust
  Scottish Centre
  Eastpark Farm
  Caerlaverock
  Dumfries DG1 4RS

British Big Cats Society
www.britishbigcats.org

South of Scotland Golden Eagle Project
c/o The Langholm Initiative
Buccleuch Mill
Glenesk Road
Langholm Dumfries and Galloway
DG 13 OES

Angling Clubs and Associations in the Scottish Borders

Border Angling Centre
97 High St.
Galashiels TD1 1RZ

*

St. Mary's Angling Club
stmarysloch@gmail.com

*

Berwick and District Angling Association
Berwick
Scottish Borders TD15 2DU

*

Whiteadder Angling Association
Duns
Scottish Borders TD11 3DW
Scotland

*

Jedforest Angling Association
Jedburgh
Scottish Borders TD8 6JY
Scotland
www.jedforest-angling.co.uk

*

Kelso and District Angling Association
Kelso
Scottish Borders
Scotland

www.kelso.bordernet.co.uk/sports/angling

*

Peeblesshire Trout Fishing Association
Peebles
Scottish Borders EH44 6JE
Scotland

*

Gatehouse and Kirkcudbright Angling Association
Kirkcudbright
Dumfries and Galloway
Scotland

*

Dumfries and Galloway Angling Association
Dumfries and Galloway DG1 4NY
Scotland

*

Upper Annandale Angling Association
Beattock
Dumfries and Galloway DG10 9PL
Scotland

*

www.wheretofish.co.uk/location/scotland/scottish-borders

STILL ON THE TRACK OF UNKNOWN ANIMALS

The Centre for Fortean Zoology, or CFZ, is a non profit-making organisation founded in 1992 with the aim of being a clearing house for information, and coordinating research into mystery animals around the world.

We also study out of place animals, rare and aberrant animal behaviour, and Zooform Phenomena; little-understood "things" that appear to be animals, but which are in fact nothing of the sort, and not even alive (at least in the way we understand the term).

Not only are we the biggest organisation of our type in the world, but - or so we like to think - we are the best. We are certainly the only truly global cryptozoological research organisation, and we carry out our investigations using a strictly scientific set of guidelines. We are expanding all the time and looking to recruit new members to help us in our research into mysterious animals and strange creatures across the globe.

Why should you join us? Because, if you are genuinely interested in trying to solve the last great mysteries of Mother Nature, there is nobody better than us with whom to do it.

We publish a journal *Animals & Men*. Each issue contains nearly 100 pages packed with news, articles, letters, research papers, field reports, and even a gossip column! The magazine is Royal Octavo in format with a full colour cover. You also have access to one of the world's largest collections of resource material dealing with cryptozoology and allied disciplines, and people from the CFZ membership regularly take part in fieldwork and expeditions around the world.

The CFZ is managed by a board of trustees, with a non-profit making trust registered with HM Government Stamp Office. The board of trustees is supported by a Permanent Directorate of full and part-time staff, and advised by a Consultancy Board of specialists - many of whom are world-renowned experts in their particular field. We have regional representatives across the UK, the USA, and many other parts of the world, and are affiliated with other organisations whose aims and protocols mirror our own.

You'll find that the people at the CFZ are friendly and approachable. We have a thriving forum on the website which is the hub of an ever-growing electronic community. You will soon find your feet. Many members of the CFZ Permanent Directorate started off as ordinary members, and now work full-time chasing monsters around the world.

Write to us, e-mail us, or telephone us. The list of future projects on the website is not exhaustive. If you have a good idea for an investigation, please tell us. We may well be able to help.

We are always looking for volunteers to join us. If you see a project that interests you, do not hesitate to get in touch with us. Under certain circumstances we can help provide funding for your trip. If you look on the future projects section of the website, you can see some of the projects that we have pencilled in for the next few years.

In 2003 and 2004 we sent three-man expeditions to Sumatra looking for Orang-Pendek - a semi-legendary bipedal ape. The same three went to Mongolia in 2005. All three members started off merely subscribers to the CFZ magazine. Next time it could be you!

We have no magic sources of income. All our funds come from donations, membership fees, and sales of our publications and merchandise. We are always looking for corporate sponsorship, and other sources of revenue. If you have any ideas for fund-raising please let us know. However, unlike other cryptozoological organisations in the past, we do not live in an intellectual ivory tower. We are not afraid to get our hands dirty, and furthermore we are not one of those organisations where the membership have to raise money so that a privileged few can go on expensive foreign trips. Our research teams, both in the UK and abroad, consist of a mixture of experienced and inexperienced personnel. We are truly a community, and work on the premise that the benefits of CFZ membership are open to all.

Reports of our investigations are published on our website as soon as they are available. Preliminary reports are posted within days of the project finishing.

Each year we publish a 200 page yearbook containing research papers and expedition reports too long to be printed in the journal. We freely circulate our information to anybody who asks for it.

We have a thriving YouTube channel, CFZtv, which has well over two hundred self-made documentaries, lecture appearances, and episodes of our monthly webTV show. We have a daily online magazine, which has over a million hits each year.

From 2000—2016 we held our annual convention - the Weird Weekend. It went on hiatus because of the illness of several of the major personnel and the eventual death of one of them. But we plan to bring it back soon. It is three days of lectures, workshops, and excursions. But most importantly it is a chance for members of the CFZ to meet each other, and to talk with the members of the permanent directorate in a relaxed and informal setting and preferably with a pint of beer in one hand. Since 2006 - the Weird Weekend has been bigger and better and held in the idyllic rural location of Woolsery in North Devon.

Since relocating to North Devon in 2005 we have become ever more closely involved with other community organisations, and we hope that this trend will continue. We have also worked closely with Police Forces across the UK as consultants for animal mutilation cases, and we intend to forge closer links with the coastguard and other community services. We want to work closely with those who regularly travel into the Bristol Channel, so that if the recent trend of exotic animal visitors to our coastal waters continues, we can be out there as soon as possible.

Apart from having been the only Fortean Zoological organisation in the world to have consistently published material on all aspects of the subject for over a decade, we have achieved the following concrete results:

• Disproved the myth relating to the headless so-called sea-serpent carcass of Durgan beach in Cornwall 1975
• Disproved the story of the 1988 puma skull of Lustleigh Cleave

- Carried out the only in-depth research ever into the mythos of the Cornish Owlman.
- Made the first records of a tropical species of lamprey
- Made the first records of a luminous cave gnat larva in Thailand
- Discovered a possible new species of British mammal - the beech marten
- In 1994-6 carried out the first archival fortean zoological survey of Hong Kong
- In the year 2000, CFZ theories were confirmed when a new species of lizard was added to the British List
- Identified the monster of Martin Mere in Lancashire as a giant wels catfish
- Expanded the known range of Armitage's skink in the Gambia by 80%
- Obtained photographic evidence of the remains of Europe's largest known pike
- Carried out the first ever in-depth study of the ninki-nanka
- Carried out the first attempt to breed Puerto Rican cave snails in captivity
- Were the first European explorers to visit the `lost valley` in Sumatra
- Published the first ever evidence for a new tribe of pygmies in Guyana
- Published the first evidence for a new species of caiman in Guyana
- Filmed unknown creatures on a monster-haunted lake in Ireland for the first time

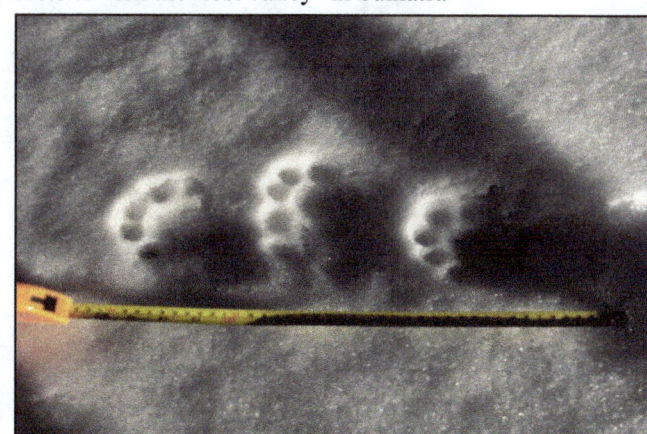

- Had a sighting of orang pendek in Sumatra in 2009
- Found leopard hair, subsequently identified by DNA analysis, from rural North Devon in 2010
- Brought back hairs which appear to be from an unknown primate in Sumatra
- Published some of the best evidence ever for the almasty in southern Russia

CFZ Expeditions and Investigations include:

- 1998 Puerto Rico, Florida, Mexico (Chupacabras)
- 1999 Nevada (Bigfoot)
- 2000 Thailand (Naga)
- 2002 Martin Mere (Giant catfish)
- 2002 Cleveland (Wallaby mutilation)
- 2003 Bolam Lake (BHM Reports)

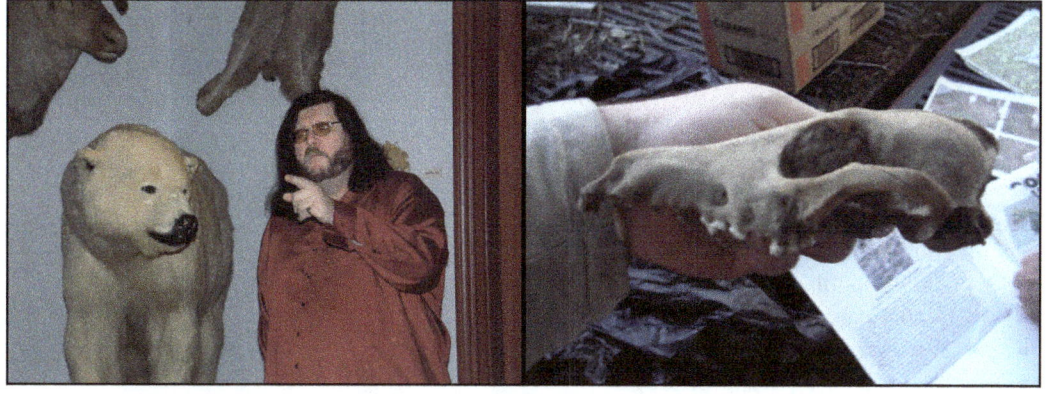

- 2003 Sumatra (Orang Pendek)
- 2003 Texas (Bigfoot; giant snapping turtles)
- 2004 Sumatra (Orang Pendek; cigau, a sabre-toothed cat)
- 2004 Illinois (Black panthers; cicada swarm)
- 2004 Texas (Mystery blue dog)
- Loch Morar (Monster)
- 2004 Puerto Rico (Chupacabras; carnivorous cave snails)
- 2005 Belize (Affiliate expedition for hairy dwarfs)
- 2005 Loch Ness (Monster)
- 2005 Mongolia (Allghoi Khorkhoi aka Mongolian death worm)

- 2006 Gambia (Gambo - Gambian sea monster , Ninki Nanka and Armitage's skink
- 2006 Llangorse Lake (Giant pike, giant eels)
- 2006 Windermere (Giant eels)
- 2007 Coniston Water (Giant eels)
- 2007 Guyana (Giant anaconda, didi, water tiger)
- 2008 Russia (Almasty)
- 2009 Sumatra (Orang pendek)
- 2009 Republic of Ireland (Lake Monster)
- 2010 Texas (Blue Dogs)
- 2010 India (Mande Burung)
- 2011 Sumatra (Orang-pendek)
- 2012 Sumatra (Orang Pendek)
- 2014 Tasmania (Thylacine)
- 2015 Tasmania (Thylacine)
- 2016 Tasmania (Thylacine)
- 2017 Tasmania (Thylacine)
- 2018 Tajikistan (Gul)
- 2020 Forest of Dean (Lynx)
- 2022 Sumatra (Orang Pendek)

For details of current membership fees, current expeditions and investigations, and voluntary posts within the CFZ that need your help, please do not hesitate to contact us.

The Centre for Fortean Zoology,
Myrtle Cottage,
Woolfardisworthy,
Bideford, North Devon
EX39 5QR

Telephone 01237 431413
Fax+44 (0)7006-074-925
eMail info@cfz.org.uk

Websites:

www.cfz.org.uk
www.weirdweekend.org

THE WORLD'S WEIRDEST PUBLISHING COMPANY

HOW TO START A PUBLISHING EMPIRE

Unlike most mainstream publishers, we have a non-commercial remit, and our mission statement claims that "we publish books because they deserve to be published, not because we think that we can make money out of them". Our motto is the Latin Tag *Pro bona causa facimus* (we do it for good reason), a slogan taken from a children's book *The Case of the Silver Egg* by the late Desmond Skirrow.

WIKIPEDIA: "The first book published was in 1988. *Take this Brother may it Serve you Well* was a guide to Beatles bootlegs by Jonathan Downes. It sold quite well, but was hampered by very poor production values, being photocopied, and held together by a plastic clip binder.

In 1988 A5 clip binders were hard to get hold of, so the publishers took A4 binders and cut them in half with a hacksaw. It now reaches surprisingly high prices second hand.

The production quality improved slightly over the years, and after 1999 all the books produced were ringbound with laminated colour covers. In 2004, however, they signed an agreement with Lightning Source, and all books are now produced perfect bound, with full colour covers."

Until 2010 all our books, the majority of which are/were on the subject of mystery animals and allied disciplines, were published by `CFZ Press`, the publishing arm of the Centre for Fortean Zoology (CFZ), and we urged our readers and followers to draw a discreet veil over the books that we published that were completely off topic to the CFZ.

However, in 2010 we decided that enough was enough and launched a second imprint, `Fortean Words` which aims to cover a wide range of non animal-related esoteric subjects. Other imprints will be launched as and when we feel like it, however the basic ethos of the company remains the same: Our job is to publish books and magazines that we feel are worth publishing, whether or not they are going to sell. Money is, after all - as my dear old Mama once told me - a rather vulgar subject, and she would be rolling in her grave if she thought that her eldest son was somehow in `trade`.

Luckily, so far our tastes have turned out not to be that rarified after all, and we have sold far more books than anyone ever thought that we would, so there is a moral in there somewhere…

Jon Downes,
Woolsery, North Devon
July 2010

CFZ PRESS

CFZ Press is our flagship imprint, featuring a wide range of intelligently written and lavishly illustrated books on cryptozoology and the quirkier aspects of Natural History.

CFZ Classics is a new venture for us. There are many seminal works that are either unavailable today, or not available with the production values which we would like to see. So, following the old adage that if you want to get something done do it yourself, this is exactly what we have done.

Desiderius Erasmus Roterodamus (b. October 18th 1466, d. July 2nd 1536) said: "When I have a little money, I buy books; and if I have any left, I buy food and clothes," and we are much the same. Only, we are in the lucky position of being able to share our books with the wider world. CFZ Classics is a conduit through which we cannot just re-issue titles which we feel still have much to offer the cryptozoological and Fortean research communities of the 21st Century, but we are adding footnotes, supplementary essays, and other material where we deem it appropriate.

http://www.cfzpublishing.co.uk/

Fortean Words is a new venture for us. The F in CFZ stands for "Fortean", after the pioneering researcher into anomalous phenomena, Charles Fort. Our Fortean Words imprint covers a whole spectrum of arcane subjects from UFOs and the paranormal to folklore and urban legends. Our authors include such Fortean luminaries as Nick Redfern, Andy Roberts, and Paul Screeton. . New authors tackling new subjects will always be encouraged, and we hope that our books will continue to be as ground-breaking and popular as ever.

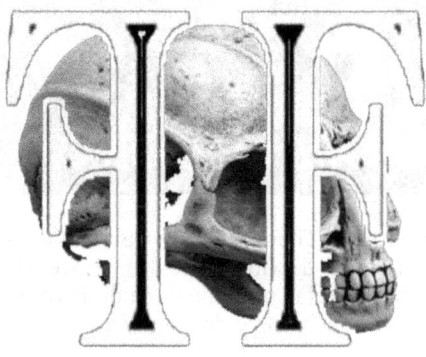

Just before Christmas 2011, we launched our third imprint, this time dedicated to - let's see if you guessed it from the title - fictional books with a Fortean or cryptozoological theme. We have published a few fictional books in the past, but now think that because of our rising reputation as publishers of quality Forteana, that a dedicated fiction imprint was the order of the day.

http://www.cfzpublishing.co.uk/

Notes

www.ingramcontent.com/pod-product-compliance
Lightning Source LLC
Chambersburg PA
CBHW071723040426
42446CB00011B/2196